PRACTICAL HISTOCHEMISTRY

PRACTICAL HISTOCHEMISTRY

J. CHAYEN

L. BITENSKY

R. G. BUTCHER

Division of Cellular Biology,
The Mathilda and Terence Kennedy Institute of Rheumatology, London

A Wiley–Interscience Publication

JOHN WILEY & SONS

London · New York · Sydney · Toronto

Library of Congress catalog card number 72–8596

ISBN 0 471 14950 0

This book is a revised version of *A Guide to Practical Histochemistry* by J. Chayen, R. G. Butcher, Lucille Bitensky and L. W. Poulter published 1969 by Oliver and Boyd Ltd., Edinburgh 1

Printed in Great Britain by John Wright and Sons Ltd. at The Stonebridge Press, Bristol BS4 5NU

ACKNOWLEDGEMENTS

We gratefully acknowledge our indebtedness to many colleagues who have contributed to our understanding and practice of this subject. They are far too numerous to be mentioned individually: many are referred to in the text; we hope the others will recognise their contributions and we hope that they will be able to approve of our use of them. But we must particularly record our deep gratitude to Mr A. A. Silcox, F.I.S.T., F.R.M.S., who originated our interest in the potentiality of cryostat microtomy for histochemistry generally, and to Dr F. P. Altman for all his help. We are very grateful to Miss M. Taylor and to Mrs S. Wiseman for their kindness in dealing with an unruly manuscript.

Over many years, and in various ways, each of us has received encouragement and support from the Medical Research Council. To the Council as a whole, and to many individuals associated with it, we offer our grateful thanks. We also gratefully acknowledge the general support of the Arthritis and Rheumatism Council for Research.

INTRODUCTION

RATIONALE OF THIS BOOK

Over the past ten years several hundred academic and applied biological and medical scientists have visited us to ask how they could apply histochemistry to their particular studies. Often their questions were purely technical: they wanted to know how to stain for a particular enzyme or substance. They saw histochemistry as an extension of histology and they wanted a simple, well established recipe; they did not want to go into the minutiae of the science underlying the staining method any more than they did when applying special stains in histology. For such enquirers we saw our task as twofold: we had to give them a reliable technique that they could follow as simply as a routine histological method but it had to be a method which, as far as we could ascertain, had a real scientific basis. Other workers came with problems in tissue metabolism and chemical dysfunction. For these, histochemistry was an extension of biochemistry. Its special advantages were that it allowed them to relate a specific activity to a particular tissue component and it was a form of biochemistry that could be done on relatively minute pieces of tissue such as could be removed safely by biopsy, or could be kept either in maintenance or proliferative culture. Some of these queries related to the mode of action of drugs, or of hormones, or of potentially toxic food additives; others were concerned with the biochemistry of disease. Such workers were primarily concerned with metabolic biochemistry; but they too did not want to delve into the minutiae of the many variants that can be found for almost every histochemical reaction (and which are discussed in detail in the specialist text-books of histochemistry such as those by Pearse and by Barka and Anderson).

Hence, over the years, we have tested a range of histochemical reactions so as to provide simple techniques which, if followed precisely, will yield reactions where the reactive compounds or enzymes are present; which will give a measure of the degree of activity actually shown in the tissue at the time of its removal; which can also indicate the total activity of which the tissue could be capable; but all of which have a rational basis. These methods were collected in a laboratory manual and they have been used by the many visitors to our laboratory. They have now been put together, in rather more detail, first of all for the many visitors who have complained that they could not copy out all the manual, and then for wider use by whoever wishes to apply histochemistry in his studies, whether as an extension of histology or of biochemistry. **No attempt has been made to include techniques of electron histochemistry. This is a special field of its own and is completely outside these terms of reference.** The main criterion we have used in selecting these methods is their reliability (for a more complete review of all methods which have been described, the reader must be referred to the comprehensive books on histochemistry); provided the instructions are followed, and provided the tissue contains the substance or enzyme to be tested, then you should obtain a visible reaction, the intensity of which should indicate the effective concentration or activity of the substance tested. We do not guarantee that these methods will give the most 'beautiful' stained preparations (although we believe that they often will); we have aimed at providing methods which will give the most beautiful preparations which are also rational and scientifically meaningful, should such meaning become required. For this is one of the hazards of histochemistry: initially a method may be selected deliberately and solely because it yields an excellent histological stain but there seem to be few histochemists who can then resist the temptation to interpret tissue metabolism in the lurid light of such staining.

WHAT IS STAINED IN HISTOCHEMISTRY?

Histochemical staining can answer two questions: it can tell you what is the gross chemical composition of cells or of a tissue, or it can tell you what is the degree of activity of the cells or tissue studied. Before launching into the methods which in fact answer these questions, it is worth considering what each of these questions implies, and consequently how the results obtained by the methods may be interpreted.

(*i*) **Histological histochemistry.** To understand the advantages of this type of histochemical enquiry, we will consider two cases which were presented to us. Pathologists were concerned over the diagnosis of a tumour in the neck of a patient. On histological examination the tissue seemed to be composed of either ganglion cells (i.e. to be diagnosed as a carotid-body tumour) or of fibroblasts (i.e. in a fibrosarcoma). Expert opinion was divided. All that was required, histochemically, was to show that the qualitative histochemistry was characteristic for one or the other cell type. Although the methods were somewhat complex (they are given in the techniques section of this book) they showed conclusively the presence of high concentrations of gangliosides, so showing that this was a carotid-body tumour. In the second case, a secondary growth was found in a lung. From the histology alone, it was difficult to be certain whether the primary growth was in the liver or adrenal. The high concentration of steroids in the cells of the secondary growth indicated strongly that the primary tumour was to be found in the adrenal.

In both these examples, histochemistry was used (in the first case with extensive controls) to provide histochemical 'stains' which indicated the chemical characteristics of the tissue and in this way histochemistry materially aided histological diagnosis. Another example of this use of histochemical 'stains' for the characteristic chemistry of a tissue is the well-known work of Willighagen on Hirschsprung's disease, which will be quoted because it involves enzyme histochemistry, not structural histochemistry as in the previous two examples. This disease is a congenital malformation which results from the absence of ganglion cells in the large bowel. Consequently that part of the bowel lacking ganglia is resected and the surgeon needs to know the upper limit of the resection. Willighagen tests the pieces of bowel that are removed and the surgeon continues resection until the latest pieces show the enzymatic activity (very high dehydrogenase activity) which characterizes these ganglion cells.

(*ii*) **Active groups and molecules.** It has often been claimed that a limitation of histochemistry is that it can show only those active groups which are *available* to the dye or reaction. In fact this need not be true, but more significantly, this is not a limitation but a major advantage. Except in the type of case quoted above, the major differences between cells lie not so much in their gross characteristics but in the degree to which their active groups or enzymes are being used. This is particularly true for different physiological and pathological conditions occurring within a single tissue or organ. The student of metabolism is well aware that cells and tissues have a great chemical or physiological reserve on which they can draw in times of stress; for such workers it is important to know to what extent the reserve is being called into play at the time of biopsy. For these studies histochemistry can show how much activity, or how many reactive groups, are manifestly available and, by suitable unmasking procedures, it can show how much (latent) activity can be made available (that is, it can give an approximation to the total gross activity).

The histochemist must therefore define his question and use the techniques to answer that specific question. The methods described in this book are designed generally to allow him to show the degree of activity manifested in the tissue at the time of biopsy; they can also yield 'total' activity if the latent reserve of activity is unmasked.

CONTENTS

PART I: THE PREPARATION OF SECTIONS

INTRODUCTION

One of the main reasons why histochemistry is bedevilled by controversy is because the tissue sections used by different workers have been prepared in very different ways. The same staining method, used on the same tissue, may yield very different results, depending solely on how the tissue and sections have been processed. It is frequently concluded, therefore, that histochemistry is an art ('histo-alchemy') and a fickle art-form at that; in fact, more often, the trouble lies in the lack of standardization of the matter to which the histochemistry is applied.

Before sections can be cut, the tissue has to be hardened and autolysis stopped. This can be done in the following ways, each of which has its peculiar limitations.

1. By chemical fixation and embedding in a hard matrix. The whole point of chemical fixation is that the fixative should react with specific active chemical groups; it must be remembered that it is just these active groups which you want to stain in histochemistry and which are the active sites of enzymes. Consequently chemical fixation cannot avoid altering the chemical—and so the histochemical—nature of the tissue. Treatment with alcohol, or with acetone, whether for fixation or as part of the process of embedding, will cause denaturation (and so also change the active chemical sites which are available for the staining reactions) and it will affect lipid–protein bonds which stabilize membranes and control enzymatic activities (e.g. as discussed by Green, 1959, and by Mazanowska *et al.*, 1966). Other fat-solvents, such as xylene or chloroform, will not only remove free fats, but will enhance the splitting of lipid–protein complexes of the protoplasm (see Lovern, 1957).

These effects and dangers of chemical fixation are well documented (e.g. Danielli, 1953; Wolman, 1955; Barka and Anderson, 1963; the effect on enzymes has been detailed by Nachlas *et al.*, 1956; Schnitka and Seligman, 1960; Holt, 1959; Holt *et al.*, 1960).

2. By freeze-drying. In this technique autolytic processes are stopped by 'quenching' tissue in liquefied gases at very low temperatures (e.g. $-190°C$); this necessitates the use of very small specimens (preferably 1–2 mm square in cross-section). Water is removed (at about $-40°C$, drying proceeding for a few days) and the dried tissue is then embedded in paraffin wax. This sets the tissue in a hard matrix so that sections can be cut with a conventional microtome. The wax is then removed (e.g. with xylene) and the tissue is then fixed, in the dried state, with absolute alcohol. The fact that the alcohol acts on dry protoplasm is said to be the reason for the improved preservation of the protoplasm over what would occur when the fresh tissue is plunged into absolute alcohol.

The drawbacks to this technique are (1) that different workers have different criteria for deciding when the tissue is 'dry' and so may obtain very different results; (2) the fact that fat solvents and alcohol must be used (and these introduce artifacts, as discussed above); (3) that the 'quenching' entails the use of liquefied gases, which are not easily available, and the use of very small specimens (and even these may show some zones which are well preserved and others which are obviously distorted). Many workers have not appreciated that a gas like nitrogen or air, when liquefied, will boil around the warm specimen and the gas so liberated will insulate the specimen from the cooling effect of the liquid gas (e.g. as was pointed out by Moline and Glenner, 1964). It is necessary to immerse the specimen in a gas which has a relatively high boiling point, like propane (or better still, a mixture of butane and propane) which has been cooled to −190°C by means of an outer bath of liquid nitrogen. These hydrocarbons have a good thermal conductivity and do not vaporize around the tissue, but they add to the technical difficulties.

This subject has been extensively reviewed (see Bell, 1956; Danielli, 1953; *Symp. Inst. Biol.*, 1952).

3. By freeze-substitution in which the tissue is quenched, as for freeze-drying, but the water is then removed by substituting methanol for the water at low temperatures (such as −70°C). The tissue can be embedded in paraffin wax, or in ester wax, provided that either it is substituted again into butanol and then taken into mixtures of butanol and paraffin wax in the oven at 60°C or if the original substitution is done in acetone and the tissue embedded in ester wax (e.g. by the method of Chayen and Gahan, 1959).

However, many of the objections discussed above apply at least as strongly to this technique.

For fuller details of freeze-substitution reference should be made to the articles of Simpson (1941); Ostrowski *et al.* (1962a, b and c); and Davies (1954).

4. By freeze-sectioning techniques. It has long been apparent that autolysis can be stopped, and tissues hardened simultaneously, by the simple expedient of freezing the specimen and cutting sections of the frozen block. The primary objection to fresh frozen sections has been the practical one: the histology is obviously distorted. The secondary objections are just as decisive even though they are more theoretical: when tissue freezes ice forms, first in the less concentrated fluid in the extracellular spaces and then even inside the cells (depending on the rate of cooling). The intracellular dehydration produced by the extracellular ice crystals increases the ionic concentration of the cytoplasm to a level which causes denaturation of the protoplasm and ruptures lipid–protein complexes (see Lovelock, 1957) whether these occur in the general cytoplasm or in cellular or subcellular membranes. During the process of cutting, the frozen tissue thaws because of the heat liberated by the impact of the block on the knife and freezes again (see Pearse, 1960, p. 21).

This process of chilling and sectioning tissue has been investigated in some detail by the present authors. They have shown that these dangers can be avoided: measurements with thermocouples under controlled conditions showed that tissue can be supercooled (i.e. no ice is formed) and this supercooled state can be maintained even during sectioning provided that this process is done at sufficiently low temperature, and with the knife cooled further to −70°C to dissipate the heat generated by the cutting (see Lynch *et al.*, 1966; Silcox *et al.*, 1965). The section is removed from the knife by apposing to it a glass slide taken from the ambient temperature of the laboratory and held by hand. The section jumps the gap between the knife (at −70°C) and the slide (at e.g. +20°C) and an imprint of the section is left as liquid water on the knife (it freezes rapidly of course). In this way the section is 'flash-dried' by distillation of the supercooled water over a temperature gradient of nearly 100°C operating over a gap of a few millimetres. Such sections are then stable.

Ice damage is undetectable in such tissue and sections. The detecting device may be a thermocouple in the tissue (as studied by Lynch *et al.*, 1966); it may be simple microscopic examination for ice crystals (or for the holes and damage produced by them, as in the study by Silcox *et al.*, 1965); it may be dark-ground illumination for investigating the degree of denaturation-aggregation of the protoplasm or it may involve a study of lipid–protein associations (as in Chayen, 1968a) or of the permeability of subcellular membranes (see Chayen, 1968b).

The actual rate at which tissue cools in the hexane bath recommended for chilling the tissue is shown in the thermocouple trace (Fig. 1). It will be seen firstly that the rate of cooling is twice as fast as that recorded by Moline and Glenner (1964) when chilling in liquid nitrogen (see remarks above); and secondly that there is no sign of ice-formation (compare with Fig. 2).

It is recommended that this procedure be used for all histochemical investigations. For certain reactions, to demonstrate particular substances, it is necessary to fix the sections specifically for the demonstration of these substances. Other substances or groups are better studied in unfixed sections. Details are given for each reaction. But the sections of tissue are standardized;

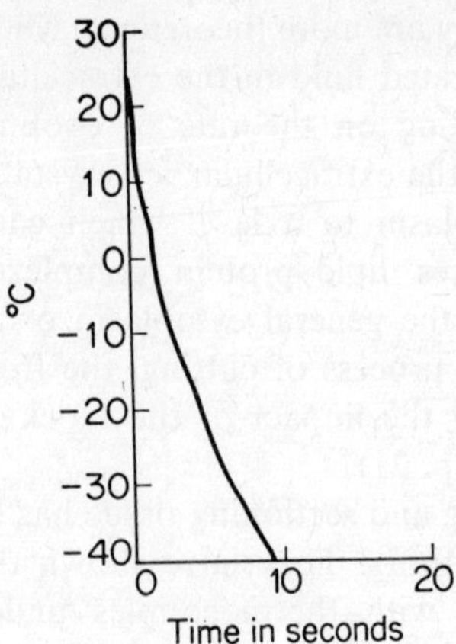

Fig. 1. The cooling curve of a piece of tissue immersed in hexane at −70°C. Note that the temperature reaches −40°C in about 9 sec and that there is no distortion (flattening) of the curve; therefore there is no indication of the formation of ice in this tissue.

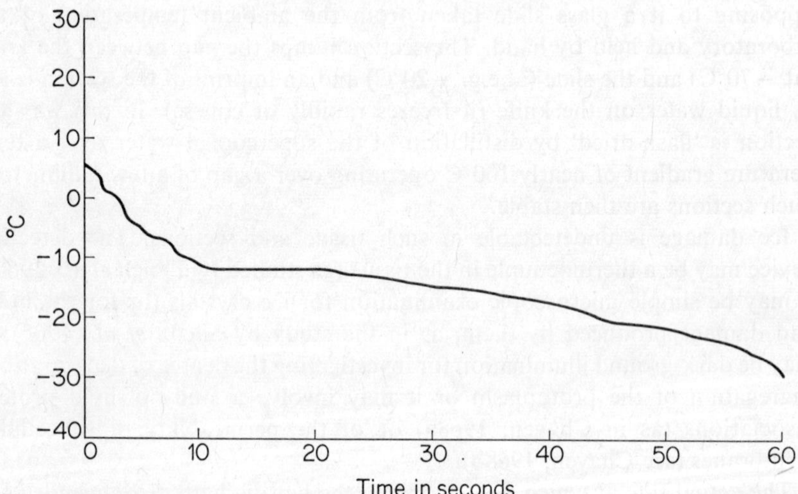

Fig. 2. The cooling curve of a piece of tissue chilled at about −20°C. In contrast to the curve in Fig. 1 there is extensive flattening of the curve, denoting ice-formation.

the preparative procedure has been shown not to introduce any apparent artifact, and the whole processing is under the rational control of the histochemist. Consequently, variations in histochemical reaction are due to true variations in activity (or in manifest activity) and so indicate metabolic differences.

METHODS

TO CHILL TISSUE

Pretreatment. One of the advantages of this procedure is that a wide range of tissues, from many animals and from plants, can all be chilled successfully by the same technique. Usually the tissue is chilled directly it is removed from the body or plant. Delicate structures, however, may benefit from immersion for 10–30 min in a 5% aqueous solution of polyvinyl alcohol (PVA; see Appendix 2) before being chilled.

Apparatus and materials. The main apparatus required is a chilling bath (a cheap polythene sandwich box is suitable if holes are cut in the lid). It should be heat-insulated by having expanded polystyrene or similar insulating matter put around it.

Also required are:
a beaker (about 50–100 ml capacity) which can be covered;
a low-temperature thermometer (recording down to -100°C);
CO_2 ice;
commercial spirit (or absolute alcohol);
hexane (B.D.H. Laboratory Reagent, boiling range 67–70°C, which must be free from aromatic hydrocarbons);
3×1 in corked specimen tubes;
Dewar flasks (the 1-gallon size is very useful for storage of tubes).

METHOD

The chilling bath is prepared by adding small chips of solid carbon dioxide (which cool quicker than do larger pieces) to alcohol in the chilling bath until a saturated solution is obtained. This state is obvious because (*i*) the alcohol–CO_2 mixture becomes viscous; (*ii*) addition of more solid CO_2 does not cause bubbling; (*iii*) the thermometer should record about -70°C.

The beaker, containing 30–50 ml of hexane, is inserted into the bath, preferably through a close-fitting space cut in the lid. More CO_2 ice is added to the bath to maintain the temperature. The temperature of the hexane should be at least -65°C before it is used. (Some workers prefer to use iso-pentane in place of hexane. We have no experience of this, but there is no reason to expect it to behave markedly differently from hexane.)

The tissue is cut into suitable pieces (which can be as large as 5 mm cubed in size) and these are dropped into the hexane at about -70°C. Small pieces (e.g. needle biopsies or material from curettage) can be blown into the hexane from a hollow spatula fitted with a large rubber bulb. Care must be taken to ensure that the tissue is plunged straight into the hexane without touching the sides of the beaker. The specimen is left in the hexane for at least 30 sec and preferably not longer than 2 min. It is then transferred, by the use of pre-cooled forceps, to a dry tube at −70°C. (*Note: from this stage onwards, the chilled tissue should be handled only with cold instruments.*) It is often advisable to shake the surface hexane off the specimen while transferring it. Care must be taken to ensure that the tissue does not warm up during this process; the specimen, in the corked dry tube, should be encased in solid carbon dioxide in the Dewar flask for storage.

A variation of this method may be used with tissue like lung which floats on the hexane. This can be pressed gently (but firmly, with a scalpel or forceps, to eliminate air-pockets) against the inside of a glass specimen tube at about −70°C (preferably low down inside the tube to avoid the heating that occurs close to the hand). The temperature of the tube can be achieved by immersing it in the CO_2–alcohol bath, or by allowing it to equilibrate with solid CO_2 packed around it in a Dewar flask (method of Chayen *et al.*, 1960; Cunningham *et al.*, 1962).

TO MOUNT THE TISSUE

This is the most hazardous operation in the whole procedure. It should be effected expeditiously, and care must be taken not to allow the tissue to be warmed at any stage.

1. Prepare another alcohol–solid CO_2 bath at about −70°C.
2. Place the metal chuck in this bath with the top clear of the alcohol; leave it to equilibrate with the bath.
3. Place a drop of water on top of the chuck. The water should begin to freeze rapidly.
4. Remove the chilled tissue from the 3 × 1 in corked tube (in which it has been stored in the Dewar flask, packed with solid CO_2) and place it in a cavity in a piece of CO_2 ice so that its orientation can be determined before it is mounted, while keeping it chilled.
5. The water on the chuck is left to freeze until there is only a thin film of water, of comparable size to that of the specimen, left unfrozen. Then, expeditiously, transfer the specimen (with cold forceps) to this film of water. The residual water will freeze the tissue to the drop of ice on the chuck.
6. Remove the chuck (with the specimen) from the bath; wipe its sides free of adherent alcohol and stand it in the refrigerated cabinet of the cryostat.

TO PREPARE THE MICROTOME FOR SECTIONING

The angle of the knife is critical. The knife is positioned as follows:

On the knife-mounting of the microtome there are four screws: two threaded into lugs at the back and two into lugs at the front. The back pair are marked into eight divisions. Before the knife is placed in the mounting the two back index screws should be unscrewed far enough to prevent them from protruding into the knife-mounting space.

The knife is then placed in the mounting and the two front screws are screwed up as tight as they will go, forcing the knife flat against the back of the mounting. The two index screws are then screwed forward until they touch the back of the knife. The number on each index screw opposite the line marked on the top of the respective lugs at the back of the mounting should then be read off and recorded.

The front screws are then unscrewed, releasing the knife. Both the index screws are then advanced one complete revolution until the same recorded number of each appears opposite the line on its respective lug. The front screws are then tightened back on to the knife so that it will be held firm and it will now be at the right angle.

Once it has been set, the knife must be cooled to $-70°C$ by packing CO_2 ice around the handle. We suggest that the knife be left, packed around with CO_2 ice for at least one hour before it is used. The ambient temperature in the cabinet must be maintained at $-30°C$ (Silcox *et al.*, 1965).

TO CUT SECTIONS

The chuck is locked on to the front of the arm of the microtome. It is positioned in such a way that the smallest edge of the block faces downwards. Blocks which have an epithelium on one edge should be positioned with this epithelium on one side and not at the top or bottom facing the knife. The screw which determines the thickness of the sections is set to the required size. In general, sections 8–10 μ thick are suitable for all histochemical work. However, tissue chilled and prepared in this way can be cut at any required thicknesses from 2–20 μ.

On the front of the microtome is the anti-roll plate assembly. As its name suggests it is designed to prevent the sections from curling up as they are cut. The new type of anti-roll plate is made of transparent Perspex and has two small nylon screws which allow you to adjust the gap between the anti-roll plate and the knife (Fig. 3). The angle of this plate to the knife is also critical but may have to be varied according to the tissue to be cut; consequently the correct angle setting will be found only by experience. However, whatever angle of the anti-roll plate is used, the top of the plate should never be lower than the cutting edge or high enough to touch the block.

TO PICK UP THE SECTIONS

The anti-roll plate keeps the sections flat against the knife while they are being cut. Then:

1. Swing the anti-roll plate away from the knife.

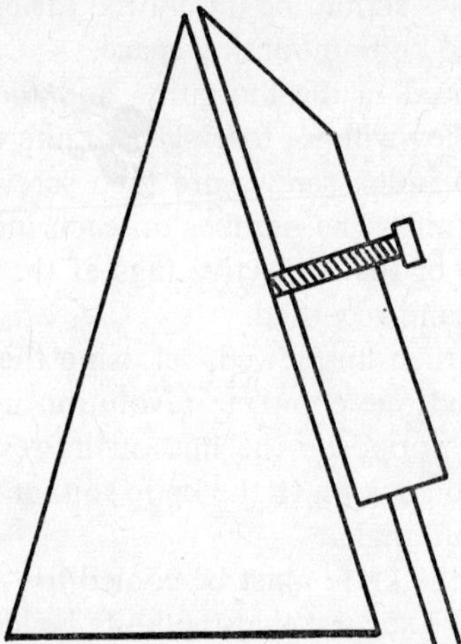

Fig. 3. Diagram of the side-view of the anti-roll plate in position against the knife, with the gap for the section adjusted by the screw.

2. Take a glass slide from the ambient temperature of the laboratory and bring it up to, and parallel to, the section on the knife. There should be no need to *press* the slide on to the section: it should move on to the slide.

3. Store the sections, on the slides, in the cryostat until used.

PART II: METHODS OF INCUBATING SECTIONS

METHODS OF INCUBATION

The simplest way is to immerse the section in a **Coplin jar**, very much as is done in conventional (but small-scale) histology. This has the following advantages:

(*a*) the solutions, in the Coplin jars, can be left in a water bath, or an incubator, or an ice bath, to equilibrate to a constant, predetermined temperature (this can be checked with a thermometer);

(*b*) the sections are immersed in a constant volume of solution, i.e. there is no danger of drying-out. Moreover since there is such a large volume of solution relative to the volume of the sections, there is no danger that the reactants will be seriously depleted during the incubation, nor that deleterious matter, produced by the sections, could become sufficiently concentrated as to have inhibitory effects on the reaction;

(*c*) it is simple to do.

The disadvantage of using Coplin jars is that they require a large volume of the incubation medium. Should this contain co-enzymes (NAD or NADP, or co-enzyme A, etc.) the cost of these substances would make histochemistry prohibitively expensive.

The usual way of overcoming this, in the past, has been to prepare only small volumes of expensive incubating media (e.g. 1 ml) and to add this as a **drop, to cover the section** which is laid horizontally in some sort of humidity chamber to stop evaporation of the drop (Fig. 4).

The advantage of this procedure is that it avoids expense. The disadvantages are:

(*a*) that frequently part of the drop does evaporate, leaving some of the section free of fluid. It is difficult to be sure that a negative reaction in part of the slide is not due to this;

(*b*) that evaporation can change the concentration of the reagents in the drop of incubation fluid;

(*c*) that the drop is not of uniform thickness over the section and so, depending on how it has spread, some parts may have a greater depth of solution above them than have others;

(*d*) that the small volume of the drop could allow inhibitory substances, produced by the histochemical reaction, to reach a concentration that could seriously affect the result of the reaction.

The **micro-cell** method (Jones, 1964) was designed to overcome the objections to the use of a drop of incubation medium. The disadvantages of this procedure are:

(*a*) it is difficult to make the micro-cells and they are awkward to use;

(*b*) the micro-cells make it virtually impossible to control the gaseous atmosphere in which the reaction is performed. The atmosphere can be controlled in both of the other methods. Thus air, or selected mixtures of oxygen, nitrogen or carbon dioxide, can be bubbled into the solution in the Coplin jar, or can be passed through the humidity chamber in which the section lies covered with a thin drop of incubation fluid.

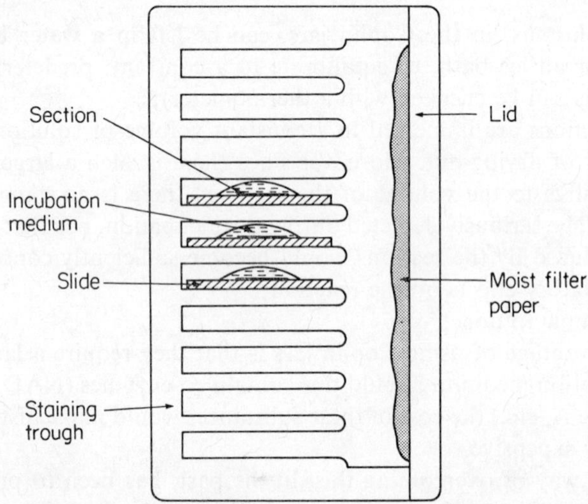

Fig. 4. Diagram of a cross-section of a humidity chamber. An ordinary staining trough stands on edge so that the sections lie horizontally. The atmosphere is kept humid by a layer of moist filter paper.

The inability to control the gaseous atmosphere (and to allow the escape of unwanted gas from the reaction) is the main reason which forced us to discard the micro-cell procedure.

The method recommended, whenever costly incubation fluid is used, is the **open-ring technique.** In this the section is picked up normally, on to a slide and brought into the open laboratory (or into a room or cabinet kept at 37°C). A Perspex ring just large enough to encircle the section, and of depth about 3 mm, is set around the section and held on to the slide by a thin film of Vaseline (Fig. 5). The incubation medium (a known volume, e.g. 0·25 ml, can be pipetted into the space bounded by the ring if this precision is required) is added to the section and confined around the section by the open ring. The gaseous atmosphere can be controlled, especially if the incubation is done in specially designed incubation boxes (Fig. 6). The rings, and the boxes, are available commercially, or can be made readily from Perspex. The whole incubation box can be put into a large incubator or into a hot room (a simple room can be constructed for £25) so that the reactions can be done

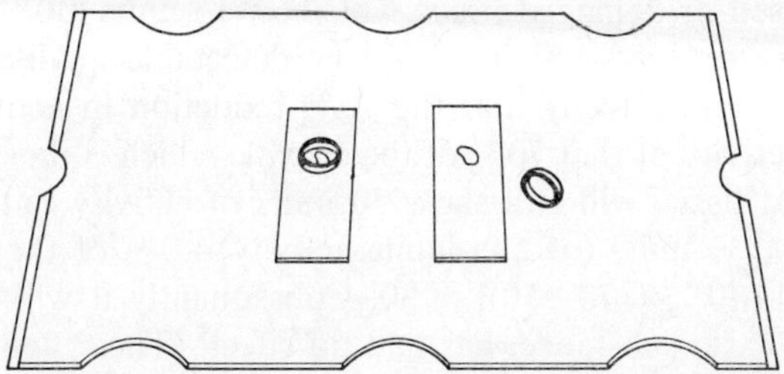

Fig. 5. Diagram to illustrate the open-ring incubation technique. On the right is a section on a glass slide with a ring of Perspex alongside just big enough to encompass the section. On the left, the ring has been held to the slide by Vaseline and can be filled with the incubation medium.

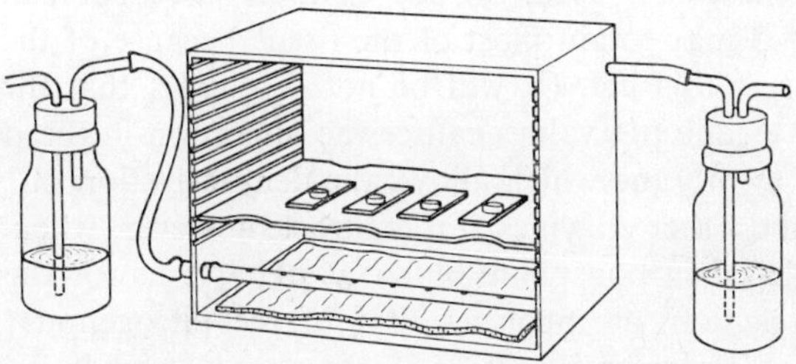

Fig. 6. Diagram of the incubation box. The slides bearing sections surrounded by Perspex rings (as in Fig. 5) are placed on Perspex trays, which can be stacked in layers. The bottom of the box is filled with moist filter paper. The front of the box is closed with a Perspex lid. The gaseous atmosphere is controlled by bubbling the appropriate gas through the wash bottle on the left; the moistened gas enters the box through the holes in the tube at the bottom of the box; the flow of gas is checked by the bubbling of the gas through the escape wash bottle on the right.

at 37°C if necessary. It should be remembered, however, that the incubation medium, and preferably the atmosphere, should be equilibrated to the incubation temperature before they are used.

THE USE OF PVA AND OF CALCIUM IONS

The dilemma of histochemists has been as follows. On the one hand, if they fix their tissues or their sections they realize that they cause such gross inactivation of active groups (and hence of enzymes) that it becomes difficult to determine whether their histochemistry has any meaning whatsoever. For example, suppose a disease, or a toxic substance (such as a food additive) or a hormone, produces 50 % reduction in the activity of a particular enzyme. It is not unusual for chemical fixation to cause 70 % loss of the activity of an enzyme. This 70 % could be 70 % of all the enzyme activity; in this case the normal tissue will show 30 units of activity (30 % left after fixation) while the diseased or damaged tissue will show 15 units (30 % of 50 %); this difference in the reduced activity might be detectable qualitatively. On the other hand, it is more likely that the 70 % reduction in activity represents 100 % loss of activity of that 70 % of the enzyme which is most labile. In this case the normal tissue will still show 30 units of activity but the abnormal tissue will show 30 units (of non-labile activity) + 0 % of the reduced labile activity, i.e. $30 + [0\% \times (70 - 50)] = 30$. Consequently it will seem as if the disease, or the toxic substance, has had no effect. (These figures are offered only to give some idea of the magnitude of changes which could be involved.)

On the other hand, if the histochemists do not fix their sections, they know that much of the undenatured section may break up, or go into solution, during incubation. Just how much will do so will depend on the pH, and on the nature of the buffer used. So, for example, a reaction done in acetate buffer at pH 4–5 may retain most of the tissue because of the 'fixing' effect of acetate ions at this pH. (It will be noted, p. 112, that this 'fixation' by acetate ions at acidic pH values causes the alteration in the permeability of the lysosomal membrane which allows the demonstration of lysosomal acid phosphatase and other enzymes.) In contrast, however, when fresh cryostat sections are placed in phosphate buffer at pH 7–8, as for the succinate dehydrogenase reaction, as much as 70 % of the nitrogenous matter of the sections goes into solution (Jones, 1965); in glycylglycine buffer only 50–60 % goes into solution (Altmann and Chayen, 1965) but this is enough to invalidate histochemistry as a scientific study of tissue metabolism and function.

This objection to histochemistry seems to have been overcome by the finding that a sufficient concentration of an inert polymer, polyvinyl alcohol (PVA), can stabilize fresh, unfixed sections even against the effect of aqueous buffer at neutral or near-neutral pH. Although this substance has been used for many years in histochemistry, the concentration which used to be recommended, namely 5–10 %, is inadequate for this purpose. It has been shown that if 20 % (w/v) of a particular polymer of PVA is added to the incubation medium for dehydrogenase enzymes, all the measurable nitrogenous matter (protein and nucleic acids) is retained in the unfixed tissue section (Altmann and Chayen, 1965). Moreover, provided that a minute amount of cytoplasmic oxidation is being done by the section, it also retains its otherwise 'soluble' dehydrogenases (Altmann and Chayen, 1966). This is effected by adding small amounts of glucose-6-phosphate, and of NADP (the co-enzyme required by the enzyme which oxidizes glucose-6-phosphate) and a hydrogen-acceptor (like neotetrazolium) to the 20 % PVA. It is interesting that the complete and functional hydrogen-transport system is essential if the minute amounts of these 'soluble' enzymes are to be retained—the presence of PVA alone is not sufficient. It is almost as if the sections have to be allowed to respire, protected by high concentrations of PVA (which is a stabilizer of colloidal systems), if the complete integrity of the section is to be maintained. It is possible that, in the future, histochemical reactions will be done mainly on sections which are allowed to respire at a low basal metabolic rate while the specific test is being done.

But this trend towards perfect integrity has its drawback as far as histochemistry is concerned. It is known (for example Lehninger, 1965) that the healthy mitochondrial membrane does not allow substrates and co-enzymes to pass from the incubation fluid into the mitochondrion. Consequently it may become impossible to demonstrate the enzyme succinate dehydrogenase in truly intact mitochondria because the substrate, succinate, cannot penetrate through the intact mitochondrial membrane and insinuate itself into the Krebs' cycle and on to the enzyme. The greater the damage done to the membrane, and to the mitochondrion, the better will be the histochemical demonstration of this enzyme activity. But if the histochemical procedure completely damages the mitochondrion (so that you get an intense histochemical stain for this enzyme) then it will be impossible to see whether disease (or drugs, or hormones) are affecting the mitochondrial membrane.

In practice this is overcome by suitable compromise. The histochemical procedure is adjusted so as to give a certain degree of mitochondrial permeability and so to give a 'reasonable' level of, e.g., succinate dehydrogenase activity. Then the effect of the potential damaging agent (or disease) is investigated. It may be that the damage will be so slight, relative to the trauma induced by the histochemical method, that no effect will be observed. In that case the compromise has not been a 'suitable' one, nor has the level of activity, permitted in the control (or normal) tissue, been 'reasonable'. It may be necessary to use the full protection of 20% PVA and of a basal cytoplasmic oxidation and to test whether the succinate dehydrogenase, when tested under these conditions, is affected by the disease or toxic agent (i.e. the 'normal' may show very low activity, because the mitochondrial membrane is intact and relatively impermeable but this permeability, and hence this enzyme activity, will increase as a consequence of the experimental, or disease, process).

This whole process of matching the histochemistry to the *function* that is to be investigated is well shown in the work on the effect of pathological and physiological conditions and of pharmacologically active substances, on lysosomes (Bitensky, 1962, 1963a and b; Bitensky and Gahan, 1962; Bitensky and Cohen, 1965; Bitensky *et al.*, 1963; Diengdoh, 1964). In these studies just sufficient histochemical trauma is used to produce entry of substrate into the normal lysosome and so allow the lysosomes to just begin to be stained by the histochemical reaction. The effect of the experimental or disease process on the stainability of the lysosomes under these controlled conditions is then examined.

A similar compromise may be required over whether calcium ions should be added to the incubation medium. These ions have the effect of gelling some protoplasm (see Heilbrunn, 1943) and they have marked effects on mitochondria (Lehninger, 1965).

The basic feature which is being emphasized in this discussion is that, slowly, histochemistry is becoming a rational science. Some of the mechanics of the subject are becoming known. Because it need not involve disruption of the tissue (as does biochemical homogenization) it can yield unique biochemical information. On the other hand, it can be used to provide histologically acceptable sections in which the stain represents the activity and localization of a certain active group or enzyme which has functional sigficance in the normal or pathological metabolism of the tissue. It may be that sometimes this type of reaction will involve deliberate alteration in the functional characteristics of the cells or of subcellular particles. As long as the fundamental mechanics of the histochemistry are understood, this deliberate alteration in cellular function (i.e. the induced histochemical trauma) can be under the complete control of the histochemist so that the stained preparation that results from it will be meaningful as well as attractive. To a considerable extent, the reactions listed in this book are designed to produce this effect without involving the histochemist unnecessarily in the detailed mechanics of the functional or metabolic effects which underlie the reactions.

PART III: METHODS OF QUANTITATION

To a surprisingly large extent every histochemist deals with quantitation. For example, very few will report a reaction as being either positive or negative; most will modify the reaction with some adjectival word or phrase, i.e. 'strongly' or 'weakly' positive; 'intense' reaction, etc. This already implies some assessment of the amount of the reaction; this assessment may be arbitrary ('a strong reaction') or it may be relative, e.g. to the reaction found in other tissues or under other experimental or physiological conditions. Particularly because the quantitative aspect is not openly recognized, the errors involved are not even considered. This whole problem was lucidly discussed by Gomori (1952) who gave the example of certain amphibians and fish which change colour very rapidly: the animal does not contain more or less pigment per unit area—it only changes the state of contraction or expansion of its pigment-containing cells. So too in histochemistry: the same amount of a tissue component or enzyme can produce a few intense granules of reaction, or an apparently strong diffuse stain; the tendency will be to say that only the latter shows a strongly positive reaction, i.e. only the latter contains much of the component studied. In fact the same actual amount of tissue component, and even of the reaction product, may be present in both cases but dispersed differently.

Some workers may be tempted to believe that electronic gadgetry may be less likely to be tricked than the human eye. Consequently they want to use a simple photocell to measure the amount of dye in a section, no matter how the dye may be dispersed. This situation was analysed by Gomori (1952). He considered the case of a microscope field which is filled by a homogeneous coloured layer which transmits only 5% of the incident light. The photocell will record, correctly, 5% transmission. He then concentrated this same amount of coloured material into discontinuous spots which filled only half the field. The photocell would record 51·25% transmission. If the same amount of coloured matter was concentrated into small, intensely coloured spots which occupied only 10% of the field, he calculated that the recorded transmission would be 90%. For here lies the real problem of photometry: the photocell records the amount of light (I) incident on it, i.e. the amount of light transmitted by the section. This has to be related to the total amount of light which would impinge on it if the section were not in the way (I_0). The microscopist then measures the extinction, or absorbance of the tissue, E, where $E = \log_{10}(I_0/I)$ (E being related inversely to the transmission, as discussed by Gomori). But the photocell does not measure the *amount of colour* in the section.

The error due to the dispersion of the dye decreases when the stain is relatively weak (and so transmits a greater proportion of the light). Gomori calculated that if the stain, uniformly distributed over the section, were to allow 80% of the light to be transmitted on to the photocell, it would transmit 82% if it were concentrated into half the field, and 91% if it occupied only 10% of the field. Hence the error in *transmission* would be relatively small but, since the absolute measure of the dye depends on $\log_{10}(I_0/I)$, the error in calculating the amount of dye present could be as great as 600%. Gomori (1952) emphasized, therefore, that the use of photocells to measure the amount of transmission through a section in which the dye is dispersed inhomogeneously can have only very limited quantitative significance. Such measurements may be of value when the dye is dispersed throughout the section, even in small discrete particles or regions, provided that each particle is only fairly weakly stained (i.e. its extinction value is up to about 0·3 or it transmits at least 50% of the light incident on it).

The very many photometric estimations of the amount of Feulgen stain for DNA or of stain, either for basic proteins or for arginine, in nuclei have succeeded mainly because the workers concerned have taken some care that the colour is fairly homogeneously distributed throughout the nucleus which they measured (for full review see Swift, 1953; Swift and Rasch, 1956; Pollister and Ornstein, 1959, who consider the limits of reliability of such measurements, as is also done in Wied and Bahr, 1970; Chayen and Denby, 1968). The development of scanning and integrating microdensitometry will be considered later.

ELUTION TECHNIQUES

The histochemist who lacks sophisticated microphotometric apparatus may still wish to quantify the amount of the histochemical reaction present in each section and so relate the amount of stain, assessed visually in the different parts, with the actual activity of an enzyme, or the actual amount of a particular substance, present in that section. Various valuable methods have been tried, notably by Defendi and Pearson (1955) and by Jardetsky and Glick (1956). The elegant micro-methods developed largely by Glick (1962, 1963) and by Lowry (e.g. 1964, Matschinsky *et al.*, 1968; Giacobini, 1969) from the micro-biochemical techniques of Lindestrøm-Lang and Holter at the Carlsberg Laboratories in Copenhagen belong more to the realm of biochemistry than histochemistry; they are basically disruptive analytical chemical methods.

The newer elution techniques for quantitative histochemistry (Altman, 1972) were derived from the procedure used in Professor Brachet's Laboratory in Brussels in 1951. Sections were treated with toluidine blue to stain the nucleic acids (as done by Davidson and Waymouth, 1944); the dye was then extracted from the section by dilute acid and estimated in a spectrophotometer. This gave a reasonable estimation of the amount of nucleic acids in the section. In a similar fashion, Defendi and Pearson (1955) performed a histochemical tetrazolium reaction for succinate dehydrogenase and then extracted the formazan from each section; the amount of enzyme activity was proportional to the amount of formazan extracted, which was measured in a spectrophotometer. Consequently in the procedure which we advocate, dehydrogenase enzymes in the section are allowed to act on their specific substrate in the presence of a tetrazole and of specific co-factor (if required); on accepting hydrogen from the dehydrogenation process the tetrazole is reduced to a coloured formazan which is precipitated at, or close to, the site at which it accepted the hydrogen. The localization of this activity can be examined by inspection of the section. [*Note*: phase-contrast microscopy is often helpful in determining the localization of the formazan in an otherwise unstained section.] By means of various elution procedures (Altman, 1969a and b; 1972) the formazans of all commonly used tetrazolium salts can be eluted and measured quantitatively, even those which bind strongly to the tissue and so yield the best localization. It is noteworthy that Altman (1972) has shown that the same order of enzyme activity can be measured by these procedures as by conventional biochemical methods, so that correctly performed enzyme histochemistry (including the use of colloid stabilizers and intermediate hydrogen-carriers) may be as precise as conventional biochemistry and involves no loss or inhibition.

ELUTION METHOD

General. Four serial sections are reacted for the particular enzyme, e.g. glucose-6-phosphate dehydrogenase, with one of the tetrazolium salts as the final hydrogen-acceptor. After the reaction, the formazan-stained sections are rinsed in distilled water and left to dry in air at laboratory temperature. (If triphenyl tetrazolium is used as the hydrogen-acceptor, it is advisable to rinse the sections very briefly in hot distilled water to avoid this formazan dissolving on prolonged immersion in cold water.) Each section is numbered and placed in a photographic enlarger. The enlarged image is projected on to white paper and the outline is traced with a pencil. The area included within this outline is then measured by means of a planimeter and the magnification factor through the photographic enlarger, at this particular setting, is determined. Thus knowing the area of the projected image (by planimetry) and the magnification of this image, the actual area of the section can be determined in square microns.

Each section, on its glass support, is then cut out from the rest of the slide, so that you have a formazan-stained section on a piece of glass which is just larger than the section. This glass fragment, with its section, is dropped into a test-tube and the minimum but measured amount of the eluant, sufficient to cover the sample, is pipetted into the tube. Usually 0·5 ml is used.

Most formazans dissolve readily with gentle shaking, into dimethylformamide (see Table 1) but some are elutable only at an alkaline pH. For these an alkaline, buffered dimethylformamide (DMF) is used. It is composed of 1 volume of buffer to 9 volumes of DMF. The composition of the buffer (Altman, 1969a) is as follows:

0·1M sodium hydroxide	95 ml
0·1M glycine + 0·1M sodium chloride	5 ml

(keep in a stoppered bottle at +4°C. The pH must be at least 11·5 and should be checked before use).

In practice it is found that, for those formazans which elute at an alkaline pH, the sections should be rinsed for a few minutes with this buffer alone before eluting in the 1 : 9 buffer : DMF solution. (Gentle warming, to 40°C in a water-bath, further assists this elution.) Moreover, in such studies, control sections should be incubated in the presence of the tetrazole alone and then subjected to this elution procedure to measure how much unreacted tetrazole is adsorbed on to the section. This is necessary only when the alkaline DMF is used because the alkalinity converts unreacted tetrazole to formazan. The difference between the amount of formazan extracted from these control sections and from the test sections (+substrate+co-factor etc.) is the amount of tetrazole reduced to formazan by enzymatic dehydrogenation of substrate and reduction of the tetrazole.

The dissolved formazan is measured in a micro-cell in a conventional spectrophotometer at the wavelength at which it absorbs maximally. The weight of formazan equivalent to the recorded absorption can be calculated readily from the data in Table 1. It is usual to measure four or five test sections (and an equal number of controls, where required) and to express the results as the mean value, with standard deviation.

The weight of formazan produced by the dehydrogenation reaction (or the amount of hydrogen, as calculated below) must then be related to the mass of the tissue. Different experimental conditions may call for different parameters of tissue mass. For example, if the water-content of a tissue doubles during the experiment, then a 10 μ section will have half the dry mass of tissue at the end of the experiment. For this reason, the best parameter is probably the nucleic acid content of tissue. This has the additional advantage that, if the polypeptide is used as the section-stabilizer, it can be estimated very simply on the same sections after the formazan has been eluted (Butcher, 1968, 1971a and b). Generally, however, it is sufficient to relate the amount of formazan to unit area, as measured by planimetry. This would be entirely satisfactory if it were certain that the thickness of the

sections were constant; then this calculation would be equivalent to the amount of formazan per unit volume. It has been found that the thickness of the sections cut by any one worker can be remarkably uniform but that the sections expand or contract according to the rate at which they are cut

TABLE 1

The elution and quantitation of various formazans (after Altman, 1972)

Formazan	*Solvent*	*Absorption maximum (nm)*	*An extinction of 1·0 is given by the following weights of the formazan (μg) in 1 ml of solvent*	*Molecular weight of the formazan*
MTT (m)	Dimethylformamide	550	21·0	335
INT (m)	Alkaline dimethylformamide	645	5·5	472
TNBT (d)	Alkaline dimethylformamide	680	11·0	839
NBT (d)	Alkaline dimethylformamide	675	9·4	749
NT (d)	Dimethylformamide	530	30·0	599
BT (d)	Dimethylformamide	575	92·0	659
TV (m)	Dimethylformamide	510	27·0	351
TT (m)	Dimethylformamide	485	21·0	301

Abbreviations: m: monotetrazole; d: ditetrazole.
TNBT: tetranitroblue tetrazolium; NBT: nitroblue tetrazolium; NT: neotetrazolium; BT: blue tetrazolium; TV: tetrazolium violet; TT: triphenyl tetrazolium.

No common names are generally used for the other tetrazoles.

(Butcher, 1971a). Uniformity can be obtained by using the automatic, motor-driven section-cutting device as is fitted to the 'automatic' Brights cryostat.

DOUBLE-BEAM RECORDING MICRODENSITOMETRY

Although the elution technique estimates the activity of the tested enzyme in the whole section, the histochemist is often more concerned with how that activity varies in the different parts of any one section. This can be achieved by using a double-beam recording microdensitometer (e.g. as manufactured by Joyce Loebl of Gateshead, England). This instrument consists of a simple microscope system in which the histochemically reacted specimen is placed on the stage which is motor-driven in a single linear scan; consequently selected regions of the tissue are passed across the objective and transmitted to a photomultiplier sequentially. The photomultiplier response is balanced against a movable wedge of known and graded density so that the wedge moves to equalize the density of the part of the specimen which is being

recorded. This movement of the wedge is recorded by a pen-recorder so that a trace of optical density is traced out corresponding to the different histological regions traversed by the scan. The linear displacement recorded by the pen can be calibrated in absolute units of density by using neutral density filters of known density (see Altman, 1971, 1972). The advantage of this technique is that, because it does not involve elution, the specimen is intact for inspection of the stain after the quantitation is complete. It was used for quantifying the dense reaction product of the 5-nucleotidase reaction in skin (Chayen *et al.*, 1971b) and has been calibrated by Altman (1971) for measuring formazan reactions.

PRECISE QUANTITATION: SCANNING AND INTEGRATING MICRODENSITOMETRY

Histochemistry has always been poorly regarded by workers in the precise disciplines mainly because of the following features:

1. The inactivation and the loss of activity due to fixation and loss in the reaction medium. These have been corrected by using controlled chilling and sectioning methods and by the use of colloid-stabilizers in the incubation medium.

2. Arbitrary reaction mixtures and lack of specificity. The methods given in this book are designed to overcome these objections.

3. The lack of quantitation.

The great advance in histochemistry in the past few years has been the demonstration that histochemical coloured reactions can be quantified, with remarkable precision, by means of scanning and integrating microdensitometers such as the Barr and Stroud GN2 and the Vickers M85. Basically these instruments are spectrophotometers which work through a microscope. You can inspect the tissue with normal white light and decide which cell-type you wish to measure. Having selected the cell, or the region of the tissue, you cut off the rest of the section by means of a stop, set at the level of the eyepiece image. A monochromating system allows you to illuminate the cells with light which is absorbed maximally by the histochemical reaction product and this image is projected on to a photomultiplier. To overcome the variation of stain even in one cell, the image is transmitted point by point, by means of an automatic scanning system which is different in the two makes of instrument. But the essential feature is that the optical inhomogeneities, which cause such errors in simple cytophotometry, are overcome by this scanning system. Each point in the selected image is measured separately and, in the extreme case, if each point is as small as the resolution of the microscope (i.e. 0·25 μ with the $\times 100$ objective) each point will be optically homogeneous when measured. The instrument integrates all the readings over the whole of the selected field and registers the integrated absorption

present in that field. The units of absorption recorded by the instrument can be converted to absolute units of extinction by measuring the absorption produced by neutral density filters of known extinction and plotting a calibration graph of absorption (as recorded by the instrument) against absolute units of extinction (from the neutral density filters). In a series of studies it has been shown that scanning and integrating microdensitometry gives the same absolute results as do the elution methods (and these, in turn, give the same activities of those dehydrogenase enzymes tested as do conventional biochemical procedures). This is a rapidly expanding area of quantitative cytochemistry and histochemistry. It has been discussed by Chayen *et al.* (1970) who have applied it to the quantitative assessment of drug-action; the subject has been reviewed in some detail, including practical aspects, by Bitensky *et al.* (1972). It has yielded valuable results in the measurement of lysosomal membrane permeability (Chayen *et al.*, 1971a) and in the estimation of dehydrogenase activity in single, selected blood cells from marrow biopsies of normal and of leukaemic patients (Stuart *et al.*, 1970; Stuart and Simpson, 1970).

CALCULATION OF THE AMOUNT OF HYDROGEN LIBERATED

For any selected tetrazole, the amount of hydrogen liberated by the reaction in the section is directly proportional to the amount of formazan produced. But 1 g (or 1 μg) of one formazan does not represent the same amount of hydrogen as does the same weight of another formazan. The reasons are as follows: when the formazan is formed, each tetrazole ring accepts two atoms of hydrogen. Consequently monotetrazoles (like MTT) accept two atoms but ditetrazoles (such as neotetrazolium or nitro-blue tetrazolium) require four atoms, i.e. two molecules of hydrogen for each molecule of the formazan. Hence 1 mole of a monoformazan represents 1 mole of hydrogen whereas 1 mole of a diformazan is equivalent to 2 moles of hydrogen. Thus it follows that to derive the amount of hydrogen (in μmoles) from a weight of a monotetrazole divide the weight (in μg) by the molecular weight of the formazan (Table 1); for a ditetrazole divide by the molecular weight but multiply by 2 (for full details see Altman, 1972).

PART IV: COMMON HISTOLOGICAL STAINS

RAPID DIAGNOSTIC SECTIONS AND STORAGE OF TISSUE

Histochemistry has a number of especial advantages. On the one hand it allows one to relate chemical activity and function to histological structure; on the other hand it gives more chemical specificity to the histology or histopathology. For the whole range of interests that lie between these extremes, it is important to be able to subject sections to some preferred histological stain. But it is quite obvious that the end-product of the histological staining method should be a section which resembles, as closely as possible, one prepared by the conventional histological procedures which include fixation, dehydration and embedding in paraffin wax. To achieve this end, some of the conventional histological techniques have had to be modified in a seemingly irrational way. However, these modifications have not only produced histological preparations which, by their conventional appearance make the histology clearer for the histochemist; they have had the additional advantage of opening new possibilities for the histopathologist. For example, it is now possible to cut a 'frozen section', for rapid surgical diagnosis; to have it stained within a few minutes of taking the specimen and yet to have a preparation which easily rivals one which would take several hours or even a few days by conventional paraffin-wax histology. The pathologist can use such sections for diagnosis just as if they were conventional preparations but he can also have additional sections cut for such histochemical procedures as may aid his work. Another drawback to the use of cryostat sections has been the difficulty of storing the 'frozen' blocks over a period of some years because in a routine pathology laboratory they must be kept should it be necessary to refer to them at some later stage. This has been overcome by the procedure by which the 'frozen' tissue can subsequently be fixed and embedded in paraffin wax; such tissue yields sections in which the histology is indistinguishable from that of similar pieces of tissue which have never been chilled but have been subjected only to fixation and embedding.

SPECIAL METHODS OF PREPARING TISSUE

Most tissues can be chilled fresh in hexane at $-70°C$ as described in Part I (p. 5). Muscle may benefit by being treated with 5% PVA for 15–30 min before it is chilled. (Sometimes the addition of small amounts of carnosine to the PVA can improve the histology of the muscle.) Pretreatment with PVA (5%, with or without about 1–2% of calcium chloride) is very advisable when delicate tissue is to be chilled: retina and necrotic malignant tissue typically benefit from such pretreatment.

Contrary to some belief, tissue which has been fixed in formalin can be chilled and sectioned just as effectively as fresh tissue. Certain precautions should be taken: the temperature of the chilling bath must be below −65°C and the temperature of the cabinet must be −30°C with the knife really well chilled by prolonged contact with solid carbon dioxide. The slides must be well albuminized (i.e. smear with egg albumin and leave to dry for 30 min before use).

FIXATIVES

Picric-formalin fixative (Bitensky *et al.*, 1963)

Formaldehyde (40% technical)	10 ml
Absolute ethyl alcohol	56 ml
Sodium chloride	0·18 g
Picric acid (wet)	0·15 g
Distilled water to make up to	100 ml

[*Note:* the picric acid is wet. It is placed on a filter paper to remove excess moisture and this picric acid is weighed.]

Picro-acetic acid fixative (Bitensky *et al.*, 1963)

As the picric-formalin, but with the addition of 5% of acetic acid.

Acetic-ethanol fixative

Either:	glacial acetic acid	10 ml
	absolute ethyl alcohol	30 ml
	Mix just prior to use.	
or:	glacial acetic acid	5 ml
	absolute ethyl alcohol	95 ml
	Mix just before required.	

Formol-calcium

Formaldehyde (40% technical)	10 ml
Calcium chloride ($CaCl_2.2H_2O$)	5 g
Distilled water to make up to	100 ml

Especially if this is to be kept as a stock solution, it is advisable to add marble chips.

Formol-saline

Formaldehyde (40% technical)	10 ml
Sodium chloride	0·85 g
Distilled water, to make up to	100 ml

Heidenhain's Susa (for cryostat sections)

Mercuric chloride	4·0 g
Sodium chloride	0·5 g
Trichloroacetic acid	2·0 g
Acetic acid (glacial)	4·0 ml
Formaldehyde (40% technical)	20 ml
Distilled water	100 ml

STAINING PROCEDURES

Normal haematoxylin-eosin method for cryostat sections. This is done at room temperature; the whole procedure takes about 45 min.

1. Remove fresh section from the cryostat and dry at room temperature for a few minutes.
2. Immerse in picric-formalin fixative (5 min).
3. Wash in 70% alcohol (10 sec).
4. Leave in periodic acid solution for 5 min (Steps 4–8 help to convert the histology to that seen in more conventional fixed and embedded sections).
5. Wash in 70% alcohol (10 sec).
6. Immerse in reducing rinse for 5 min.
7. Wash in 70% alcohol (10 sec).
8. Leave in absolute alcohol for 1 min.
9. Stain in Ehrlich's haematoxylin for 20 min.
10. Rinse in tap water.
11. Differentiate in acid-alcohol (1% concentrated hydrochloric acid in 70% alcohol) according to taste (e.g. up to 10 sec).
12. 'Blue' in alkaline water (5% sodium bicarbonate solution) for about 30 sec.
13. Stain in a 0·5% aqueous solution of eosin for 40–60 sec, according to taste.
14. Rinse briefly in tap water.
15. Dehydrate through a graded series of alcohols, clear in xylol and mount in DePeX.

Extra solutions required for this method

Periodic acid solution
Dissolve 0·4 g of periodic acid in 15 ml of distilled water. To this add 35 ml of absolute alcohol in which is dissolved 0·135 g of hydrated crystalline sodium acetate. The final solution should be stored in the dark.

Reducing rinse
To 20 ml of distilled water add 1 g of potassium iodide and 1 g of sodium thiosulphate. Add, stirring continuously, 30 ml of absolute alcohol and then

0·5 ml of 2 N hydrochloric acid. A precipitate of sulphur will form; this is allowed to settle.

Modified haematoxylin-eosin method (Roberts, 1966). The special advantages of this procedure are that it enhances the clarity of the nuclei and is quicker than the normal technique in that it takes about 15 min to complete.

All stages can be done at room temperature.

1. Take the section straight from the cryostat cabinet. Dry it in air for 5 min, preferably with slight warming.
2. Fix in picro-acetate fixative (3 min).
3. Wash in 70% alcohol (10 sec).
4. Immerse in celestine blue solution for 3 min.
5. Rinse briefly in tap water.
6. Stain with Mayer's haemalum for 3 min.
7. Rinse in tap water.
8. Differentiate to taste in acid-alcohol (1% concentrated hydrochloric acid in 70% alcohol).
9. 'Blue' in alkaline water (containing a few drops of a saturated solution of lithium carbonate).
10. Stain with a 0·5% aqueous solution of eosin (10 sec).
11. Rinse briefly in tap water.
12. Dehydrate through a graded series of alcohols; clear in xylol and mount in DePeX.

Extra solutions required for this method

Celestine blue solution
Dissolve 2·5 g of iron alum in 50 ml of distilled water. Add 0·25 g of celestine blue; boil for 3 min; filter when the solution is cool.

Mayer's haemalum
Dissolve 1 g of haematoxylin in 1 litre of distilled water (heat if necessary). Add 50 g of ammonium alum; shake to dissolve. Then add 0·2 g of sodium iodate, 1 g of citric acid and 50 g of chloral hydrate.

Van Gieson stain for collagen

1. Take the fresh section from the cryostat cabinet; dry it in air for 5 min.
2. Fix either in picrate-formalin for 15 min or in 1 : 3 acetic-alcohol for 10 min.
3. Rinse in tap water.
4. Stain with celestine blue for 3 min.
5. Rinse briefly in tap water.
6. Stain in Mayer's haemalum for 3 min.
7. Rinse briefly in tap water.

8. Differentiate very briefly (if necessary) in 1% acid alcohol. (*Note*: further differentiation occurs during step 12 owing to the picric acid in the Van Gieson solution.)

9. Wash in tap water.

10. 'Blue' in 5% aqueous solution of sodium bicarbonate.

11. Wash in distilled water.

12. Stain with the Van Gieson solution for ½–1 min.

13. Blot dry.

14. Dehydrate in absolute ethyl alcohol, clear in xylol and mount in DePeX.

Result

Nuclei—blue or black.
Collagen fibres—red.
Muscle and other tissue—yellow.

Solutions required for this procedure

Celestine blue, Mayer's haemalum and acid alcohol as for the modified haematoxylin-eosin method (p. 25).

Additional solutions

Van Gieson's solution

1% aqueous solution of acid fuchsin	10 ml
Saturated aqueous solution of picric acid	90 ml
Distilled water	100 ml

Boil for 3 min.

Gordon and Sweet's silver impregnation method for reticulin

1. Take the fresh section from the cryostat cabinet and dry it thoroughly in air.

2. Fix in picric-formalin for 5 min.

3. Rinse in distilled water.

4. To oxidize the tissue, treat for 1–5 min in the acidified permanganate solution.

5. Rinse in distilled water.

6. Bleach until white with 1% solution of oxalic acid.

7. Wash well in several changes of distilled water.

8. Mordant by immersing for 2–15 min in 2% aqueous iron alum.

9. Wash well in several changes of distilled water.

10. Treat for 5–8 sec in Wilder's silver bath.

11. Rinse thoroughly in distilled water.

12. Reduce in formol-calcium (neutral 10% formaldehyde) for 10–30 sec.

13. Rinse in tap water.
14. 'Fix' the silver in 5% solution of sodium thiosulphate (2–5 min).
15. Wash well in tap water.
16. Dehydrate through a graded series of alcohols, clear in xylol and mount in DePeX.

Result

Reticulin fibres—jet black.
Collagen fibres—golden brown.

Solutions required for this procedure

Acidified permanganate solution

0·5% potassium permanganate in water	47·5 ml
3% sulphuric acid	2·5 ml

Wilder's silver bath

To 5 ml of a 10% solution of silver nitrate in water add ammonia ('880') drop by drop. A precipitate will form. Continue to add the ammonia, drop by drop, until the precipitate is just redissolved. Then add 5 ml of a 3% solution of sodium hydroxide. Again add ammonia, drop by drop, until the solution just becomes clear. Add distilled water to make the final volume 50 ml.

This solution should be stored in a dark bottle and will keep for several months.

The phosphotungstic acid-haematoxylin method (we are indebted to Miss S. Darracott for this procedure, as applied routinely to cryostat sections).

1. Take the section from the cryostat cabinet. Dry the section in air.
2. Fix in picric-formalin (5 min).
3. Rinse in water.
4. Treat with Lugol's iodine (2–3 min).
5. Rinse in distilled water.
6. Treat with a 5% solution of sodium thiosulphate (2–3 min).
7. Rinse in distilled water.
8. Immerse in acidified potassium permanganate (2–3 min).
9. Rinse in tap water.
10. Bleach in 1% oxalic acid (until white).
11. Wash for 5 min in running water.
12. Leave in the PTAH solution (overnight).
13. Shake off the excess fluid but do not wash.
14. Place in 70% alcohol (2 min).
15. Dehydrate; clear in xylol; mount in DePeX.

Result. This reaction is particularly good for identifying isolated myoblasts and for cross-striations in striated muscle. It is also good for cellular detail generally.

Solutions required for this method

Lugol's iodine (see p. 51).

Acidified potassium permanganate as for Gordon and Sweet method (p. 27).

PTAH solution
Dissolve 0·1 g of haematoxylin in distilled water (up to 50 ml). Also dissolve 2·0 g of phosphotungstic acid in distilled water (again up to 50 ml may be used). Mix. Make up the final volume to 100 ml. Leave to ripen for 3 months.

Rapid ripening can be achieved by adding 0·017 g of potassium permanganate to the solution.

[*Note:* it may not be widely appreciated that isolated myoblasts or smooth muscle can be identified, even in sections of paraffin-embedded tissue which have been stained with haematoxylin and eosin, and mounted in DePeX. To do this, polarized light is used: muscle appears white against a darker background. This does not require a complete polarized light microscope: sheets of polaroid below the condenser and in or above the eyepiece often suffice. Difficulty may be met if other mounting media are used. For details of polarized light see Chayen and Denby, 1968.]

Iron haematoxylin

1. Fix dried cryostat sections in picric-formalin (5 min).
2. Wash with water.
3. Immerse in iron alum solution (5% ferric ammonium sulphate) for 30–45 min at 56°C.* This step mordants the tissue.
4. Wash well in distilled water.
5. Stain in the haematoxylin solution at 56°C for 30–45 min.*
6. Wash well in distilled water.
7. Differentiate in a 2% solution of ferric ammonium sulphate to taste, i.e. until the particular structures to be studied are clearly defined.
8. Wash in running water for 5 min.
9. Dehydrate through a graded series of alcohols; clear in xylol and mount in DePeX. (Some structures can be seen more clearly if the section is dehydrated and mounted in Euparal.)

The haematoxylin solution

Haematoxylin	0·5 g
Absolute ethyl alcohol	10 ml
Distilled water	90 ml

*Alternatively these stages can be done at room temperature but they then take 12–24 hr.

Allow to ripen for 4–5 weeks. Alternatively add 0·1 g of sodium iodate to ripen immediately.

The methyl violet method for the demonstration of amyloid

1. Dry a fresh cryostat section in air.
2. Immerse in a 1% aqueous solution of methyl violet (2–5 min).
3. Rinse in distilled water.
4. Differentiate in 0·5–1% acetic acid until the amyloid is pink and the rest of the tissue remains blue-violet.
5. Wash well in distilled water.
6. Mount in Farrants' medium.

To store 'frozen' blocks in paraffin wax. After fresh sections for histology and histochemistry have been cut from correctly chilled tissue, the block of tissue can be treated as follows:

1. Remove the tissue from the block-holder and put into a refrigerator at +4°C. Leave for 1 hr.
2. Immerse in formol-saline at room temperature. Fixation may take at least 24 hr.
3. Dehydrate, clear and embed in paraffin wax by whatever method is preferred for normally fixed tissue.

These paraffin blocks can be treated in exactly the same way as those of tissue treated by conventional histological procedures. When required they can be sectioned, dewaxed, stained by the usual histological methods and should yield preparations indistinguishable from tissues which have never been chilled.

PART V: THE ANALYSIS OF CHEMICAL COMPONENTS OF CELLS AND TISSUES

GENERAL ANALYSIS

Histochemistry should be able to analyse the chemical composition of parts of cells. As an academic exercise this has been of interest in studies on cell growth and differentiation: e.g. McLeish (1959) measured the increase in DNA and in arginine (and hence in histone-like protein) in nuclei of growing plant cells; Sandritter and co-workers measured similar changes in normal and malignant cells growing in tissue culture (e.g. Sandritter and Krygier, 1959; Sandritter and Fischer, 1962; Sandritter and Kleinhans, 1964) and Richards (1960) correlated changes in amounts of DNA and of protein in nuclei during cell division.

As a practical, applied problem this becomes more complex. The pathologist may see abnormal matter either inside or outside cells and be concerned to know of what it is composed: is it a mass of dead cells (which will contain nucleic acids and proteins); is it an abnormal protein complex, e.g. 'fibrinoid'; is cholesterol present or does it contain calcium? Generally the histologist applies a number of tests, selected largely at random. However in this Part we will suggest that this type of problem can be investigated by a rational sequence of tests, very much as the analytical chemist applies the sequence of 'group separation tests' to decide the chemical composition of an inorganic sample.

Material. Sections of unfixed tissue. Serial sections will be required so that the tests can be applied, in sequence, to the same matter.

STAGE 1 (Fig. 7): The determination of the major classes of compound present

In this stage the major classes of compound present in the material are decided:

(*i*) Test for **protein** by the DNFB or by the tetra-azotized dianisidine method (or by both of them used separately). If positive, the nature of the protein will be examined in STAGE 2.

(*ii*) Test for **acidic groups,** such as nucleic acids. At this stage all that is required is to stain with toluidine blue (at pH 6) or with the methyl green-pyronin mixture (at pH 4·2). If positive then tests in STAGE 3 will have to be done to decide on the nature of the acidic matter present.

(*iii*) Test for **carbohydrate** by the use of a controlled periodic acid Schiff test. If positive, tests in STAGE 4 will have to be used.

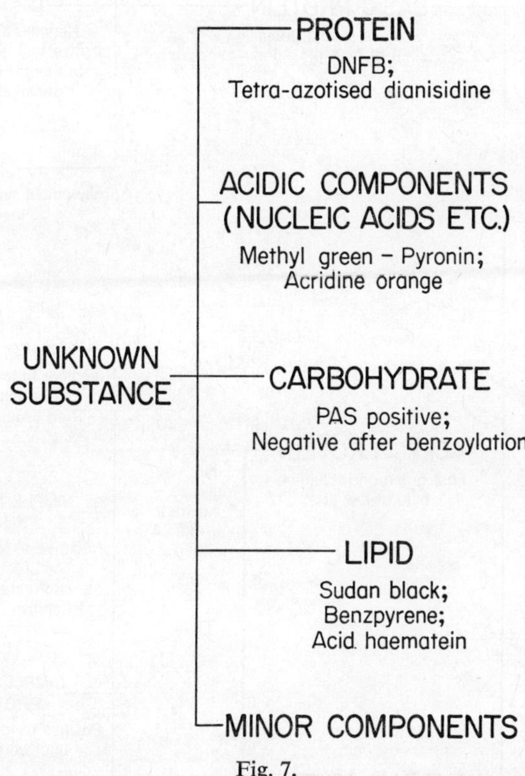

Fig. 7.

(*iv*) Investigate the material for the presence of **lipid.** Sudan black is a good general indicator of lipid although benzpyrene may be rather better, both as regards sensitivity and because it does not involve the use of a fat-solvent. If positive, apply the methods of STAGE 5.

(*v*) When all these have been tested (*i–iv* inclusive, above), and when the major components have been defined (stages 2–5 inclusive, below) **minor components** can be examined, as described in STAGE 6.

STAGE 2 (Fig. 8)

Proceed with this stage only if a positive reaction is obtained with DNFB and/or tetra-azotized dianisidine method and/or Baker's tyrosine technique.

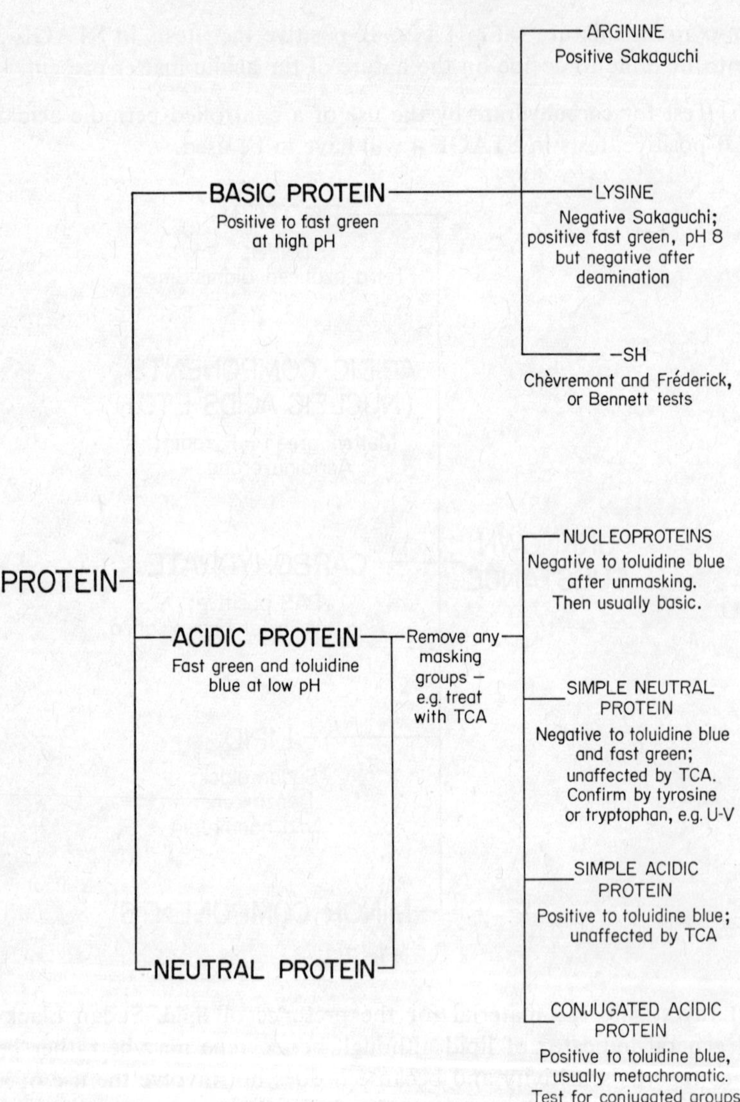

Fig. 8.

If negative to all of these it is unlikely to contain protein (and so omit this stage). If positive, then first decide whether the protein is predominantly basic, acidic or neutral:

2a It is basic if it stains with acidic dyes (such as fast green) **at a high pH value.** (*Note*: above pH 7 the —NH_2 groups present in all proteins are relatively un-ionized and so cannot bind the acidic dye whereas the more basic groups remain ionized and so stain with fast green.)

2b It is acidic if it stains with basic dyes (such as toluidine blue) **at low pH values.** (*Note*: below pH 6 the carboxyl groups present in proteins are relatively un-ionized so that if the material stains at pH 4·2 it may be taken that it contains conjugated acidic matter such as the nucleic acids or acidic polysaccharide).

2c It is a neutral protein if it shows neither of these stain characteristics, i.e. it may stain with both fast green and with toluidine blue at about neutral pH but staining with the basic dye is suppressed at more acidic pH values while staining with the acidic fast green is suppressed the more basic is the pH. (Methods of estimating the pK of protein by such procedures are well discussed by Levine, 1939).

If the protein appears to be basic this is probably due to its containing a high proportion of basic amino-acids. Therefore:

2d Test for arginine by the modified Sakaguchi reaction.

Test for lysine; lysine can be assumed to be present when the protein stains strongly with fast green at pH 8 but is negative after deamination, provided that the Sakaguchi reaction (for arginine which will respond similarly) is either negative or weak.

Test for —SH groups by the method of Chèvremont and Fréderick or by Bennett's mercury orange method.

2e If the protein appears to be acidic this could be due to its containing a high proportion of acidic amino-acids like glutamic acid (whose ionization is suppressed at pH 4·2); however it is more probably due to its being a neutral (or even a basic) protein which is conjugated to an acidic compound. Hence both neutral and 'acidic' proteins can be tested together, as follows:

Treat the section with a 5% aqueous solution of trichloroacetic acid (TCA) at 90°C for 15 min. This will remove the nucleic acids. Then:

Simple neutral protein will stain with fast green and with toluidine blue only close to neutrality (as before); it will be unaffected by the TCA. It is advisable to prove that such material is proteinaceous by testing for tyrosine or for tryptophan, as described later in this chapter.

Simple acidic protein will stain positively with basic dyes, like toluidine blue, but not appreciably with acidic dyes (like fast green) at pH values of between 4·2 and 6 as before; their staining will not have been affected by the TCA (although some intensification may be observed because of the coagulating effect of the TCA and the unmasking of acidic side-chains due to the denaturation of the protein).

2f Nucleoproteins will have stained strongly and orthochromatically (blue) with toluidine blue at pH 4·2 and with the methyl green-pyronin method before the TCA treatment; after TCA extraction these reactions will be abolished and the residual protein will stain either as a neutral or as a basic protein (e.g. the latter will be stained by fast green at pH 8). If it behaves as a basic protein then it may be subjected to the tests for basic amino-acids detailed above.

2g Other conjugated acidic proteins may be either lipo-protein or protein conjugated with acidic polysaccharide. If the former, it will behave as a simple protein and will also stain for lipid (see STAGE 5). If conjugated to acidic polysaccharide it will stain positively—and usually metachromatically (purple to red) with toluidine blue and this property will not have been lost as a result of the TCA treatment. It should then be tested for the acidic conjugate as in STAGE 4 (for polysaccharide) or STAGE 3 (for other acidic groups such as sulphate or phosphate).

STAGE 3 (Fig. 9)

Acidic substances have been shown to be present (as discussed in STAGE 1). In this stage, their nature is determined:

3a Stain with methyl green-pyronin. If green, or blue-green **test for DNA** by the Feulgen reaction; or by Kurnick's methyl green method (or by ultra-violet microscopy, if available). If these tests yield apparently positive results, use controlled digestion with crystalline deoxyribonuclease to confirm that DNA is indeed present.

3b If DNA is not present, or if it is not the sole constituent, all that is known is that acidic matter is present, i.e. it stains at pH 4·2 with basic dyes like pyronin (in the methyl green-pyronin test) or toluidine blue.

3c Test for RNA by observing if the basophilia is decreased after controlled digestion with ribonuclease. (If ultraviolet microscopy is available, the maximal absorption in the 265 mμ region, due to the purines and pyrimidines, is helpful in proving that it is a nucleic acid.)

3d If the basophilia is not removed by digestion with ribonuclease and deoxyribonuclease, tests for **other acidic groups** must be used. These tests should also be applied where the nucleases are not available. (*Note*: RNA can usually be removed almost as effectively as by ribonuclease by immersing the section in 1 N hydrochloric acid at 60°C for 6 min.)

3e Stain with toluidine blue at pH 1. Only **phosphate** or **sulphate** is likely to remain ionized at this pH and so to be capable of staining.

3f Apply Ebel's test for phosphate and polyphosphate. If negative, the acidic group is probably sulphate. Usually such substances stain metachromatically (pink or red) with toluidine blue.

3g If the material stains with toluidine blue at pH 4·2 but the staining is abolished at pH 1, it is probably due to acidic groups of the type found in **acidic** (but not sulphated) **polysaccharides**. The ability of such groups to stain with basic dyes is lost after decarboxylation; they should be positive with the periodic acid–Schiff method.

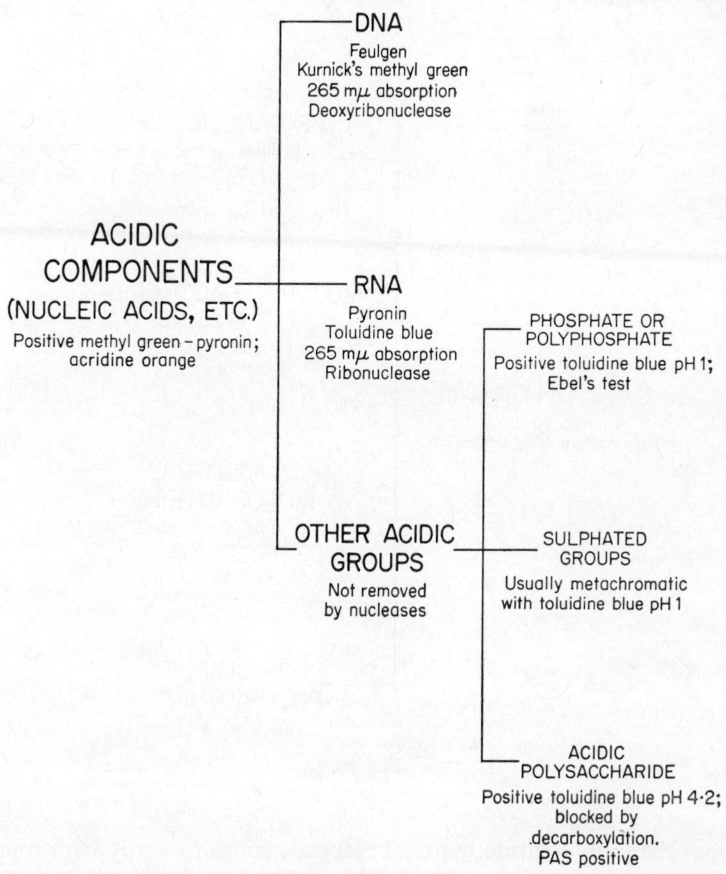

Fig. 9.

STAGE 4 (Fig. 10)

The material is positive with the alcoholic periodic acid–Schiff (PAS) test, but is negative (or the reaction is markedly depressed) after benzoylation. Hence it contains **carbohydrate.**

Is it removed by rinsing in water or in dilute acid (e.g. 0·1 N trichloroacetic acid)? If it is lost after this treatment it is either free carbohydrate or a simple polymer such as glycogen or starch.

4a (*i*) Test with Lugol's iodine for **starch**.
(*ii*) Test with Best's carmine for **glycogen**. Or use the alcoholic PAS test both with and without prior treatment with diastase or with α and β amylases.

4b If it is not removed by dilute acid it may be linked to protein. Hence apply a test for protein (STAGE 2).

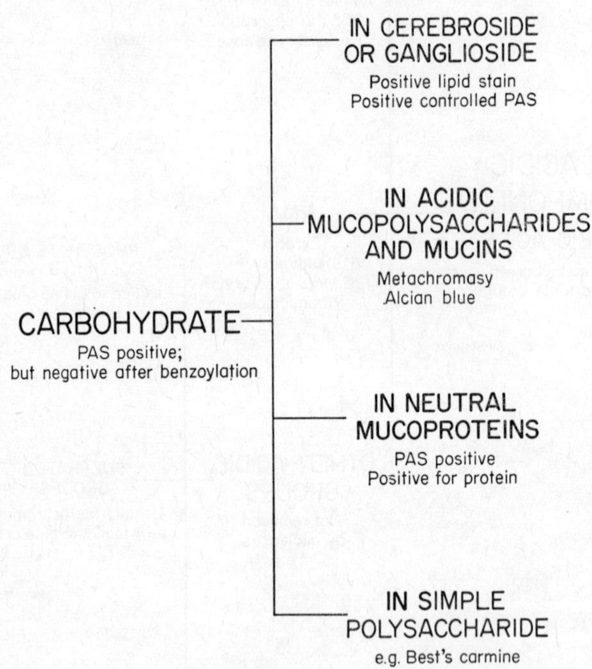

Fig. 10.

4c Does the carbohydrate material react metachromatically with toluidine blue? If so, it may be an **acidic polysaccharide**, also possibly linked with protein. Test with Alcian blue for acidic mucopolysaccharides or with Stoward's reactions to confirm this.

4d The carbohydrate material may comprise part of a **cerebroside** or **ganglioside** complex. These will be positive with lipid stains (STAGE 5). Moreover the presence of unsaturated fatty acids in such complexes may give an apparent PAS-reaction which will not be abolished by benzoylation but which will require oxidation (e.g. by bromination; this will 'saturate' the fatty acids) prior to treatment with periodic acid.

STAGE 5 (Fig. 11)

The material contains **lipid** if it stains with benzpyrene or with Sudan black (although the 'burnt' Sudan black method or some other unmasking technique may have to be employed).

5a **Steroids** may not respond well to such lipid colourants. Consequently test for autofluorescence or for birefringence. Adam's reaction, or the sulphuric acid method, may help to demonstrate steroids.

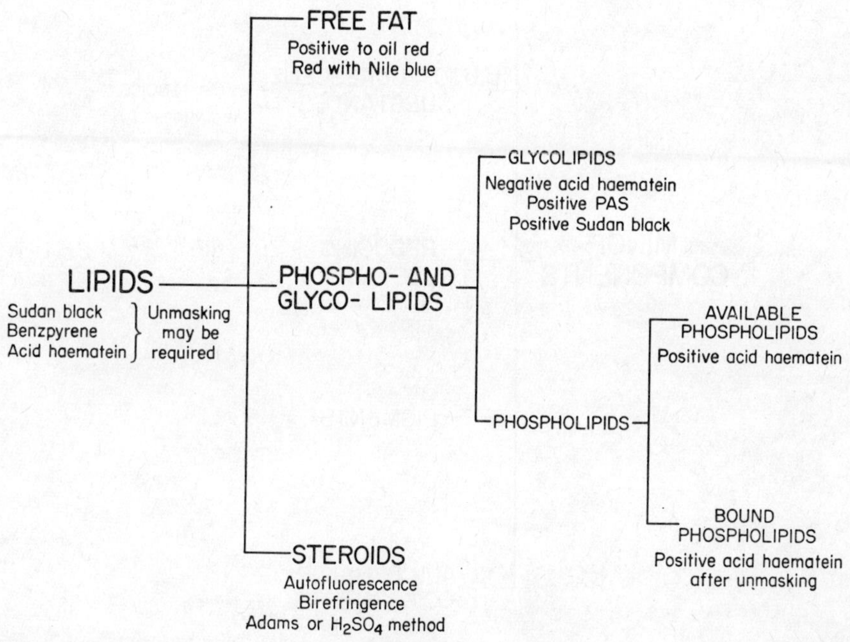

Fig. 11.

5b **Free fat** will take up lipid colourants very readily, even those with lower affinity for lipids than is possessed by Sudan black (which also stains phospholipids which are not as 'fatty' as are free triglycerides). Hence test with oil red. Moreover neutral triglycerides should yield a red colour with Nile blue.

5c Test with the acid haematein method: If positive (but negative after bromination), then the material contains **phospholipid**. The extent to which the phospholipid is masked can be tested by the unmasking procedure (below).

5d If the material stains with benzpyrene or Sudan black (i.e. it contains lipid) but is negative to oil red (hence not freely available triglyceride) and to the acid haematein test, even after unmasking procedures (i.e. it does not contain phospholipid) and it gives a positive PAS reaction (which is lost after benzoylation) then it contains a **glycolipid.**

STAGE 6 (Fig. 12)

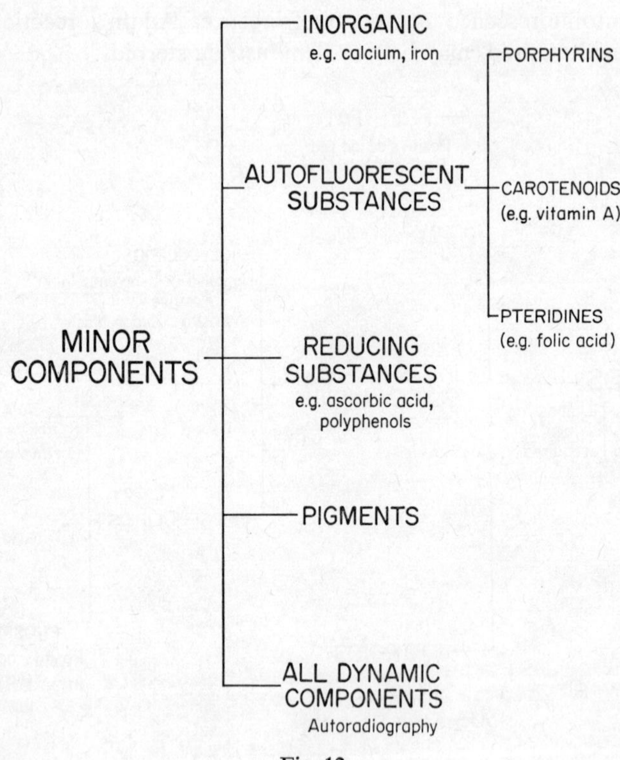

Fig. 12.

PRACTICAL METHODS

STAGE 2: REACTIONS FOR PROTEIN

DINITRO-FLUORO-BENZENE METHOD (see Danielli, 1953)

Rationale. Dinitro-fluoro-benzene (DNFB) reacts with amino-acid residues and groups such as tyrosine, $—NH_2$ and —SH (Sanger, 1945). The colour produced is too weak for histochemistry. It is intensified by synthesizing an azo dye from the DNFB–protein complex inside the section.

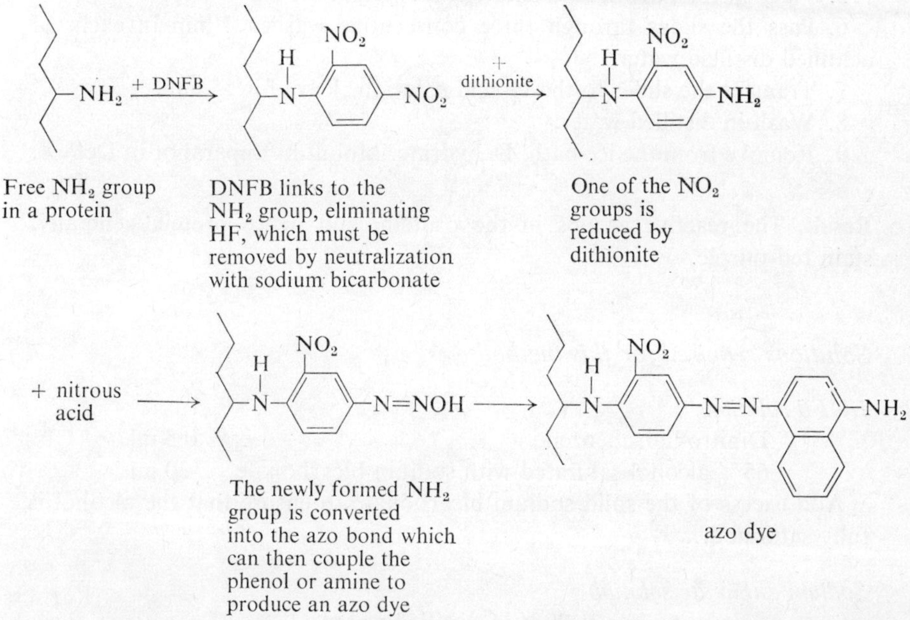

NH_2 + DNFB → NO_2, H, N, NO_2 + dithionite → NO_2, H, N, NH_2

Free NH_2 group in a protein

DNFB links to the NH_2 group, eliminating HF, which must be removed by neutralization with sodium bicarbonate

One of the NO_2 groups is reduced by dithionite

+ nitrous acid → NO_2, H, N, N=NOH → NO_2, H, N, N=N, NH_2

The newly formed NH_2 group is converted into the azo bond which can then couple the phenol or amine to produce an azo dye

azo dye

Application. Since the DNFB will react with groups which must occur in protein, it can be used as a general protein stain. It can probably be made specific for a particular group by blocking the other groups which can react. For example, Danielli (1953) suggested that the DNFB will react with tyrosine alone if $—NH_2$ groups are blocked by prior treatment with nitrous acid, and —SH groups by treatment with iodoacetamide.

METHOD

The sections need not be fixed for this procedure. [If fixation is required (to free some of the reactive groups which otherwise might be masked, e.g. by nucleic acid) then fix them in 5% acetic acid in absolute ethanol for 10 min. Take the sections through graded concentrations of alcohol to water before proceeding to 1.]

1. Place sections for 2 hr in the DNFB solution, in a Coplin jar. Stir occasionally to maintain the saturated condition of the bicarbonate solution.
2. Wash in 60% alcohol (to remove adsorbed DNFB).
3. Place in 20% solution of sodium dithionite at 37°C for 10 min.
   [*Note:* the solution should be in a Coplin jar and at 37°C before the sections are immersed.]
4. Wash in distilled water at room temperature.

STAGES 5–8 INCLUSIVE SHOULD BE DONE IN AN ICE BATH (and the solutions should be stood in the bath to cool before they are needed).

5. Place in the nitrous acid solution for 5 min.
6. Pass the slides through three consecutive washes, 1 min in each, of acidified distilled water.
7. Transfer the slides to the H-acid solution: leave for 15 min.
8. Wash in distilled water.
9. Remove from the ice bath. Dehydrate. Mount in Euparal or in DePeX.

Result. The reactive groups in the protein, and hence protein generally, stain red-purple.

Solutions required for this method

DNFB solution

Dinitrofluorobenzene	0·5 ml
65% alcohol saturated with sodium bicarbonate	30 ml

Add excess of the solid sodium bicarbonate to ensure that the alcohol is fully saturated.

Sodium dithionite solution

20% in distilled water

[*Note:* dissolve in warm water (e.g. at 40°C).]

Nitrous acid

2% solution sodium nitrite	1 volume
0·1 N hydrochloric acid	1 volume

Mix just before use.

H-acid solution

H acid (8-amino-1-naphthol-3-6-disulphonic acid)	0·25 g
1% sodium bicarbonate solution	50 ml

TETRA-AZOTIZED DIANISIDINE METHOD (see Danielli, 1953)

Rationale. Diazonium hydroxides react with histidine, with tryptophan and with tyrosine (also with many phenols), yielding a coloured complex. This colour can be intensified by coupling another phenol or amine on to the unreacted diazo-group of the diazonium hydroxide. Of the various diazonium hydroxides that are available, probably the most suitable is the tetra-azotized dianisidine.

—OH + tetra-azotized dianisidine → —OH, N=N— —N=N—OH

phenolic side group (such as tyrosine) on peptide chain of protein

tetra-azotized dianisidine linked to the phenolic group of the protein (usually weakly coloured)

→ —OH, N=N— —N=N— —NH_2

The colour is intensified by adding a synthetic amine or phenol which couples with the free diazo bond. There is no reason why the terminal —NH_2 should not be converted into a diazo group by treatment with nitrous acid, and be allowed to bind yet another molecule of the synthetic amine, as was done in the DNFB method. This would intensify the colour still further, if required

Application. Since the tetra-azotized dianisidine reacts with tyrosine as well as with histidine and tryptophan, it acts as a useful general stain for protein. It can be made specific for one or other of these amino acids by pretreating the sections with substances which will block the reactive groups of the other amino acids. According to Danielli (1953) pretreatment with dinitrofluorobenzene (DNFB) will block tyrosine and probably histidine so that any reaction with tetra-azotized dianisidine would then be due to tryptophan alone. Pretreatment with performic acid should eliminate tryptophan and leave only histidine and tyrosine available for the tetra-azotized dianisidine reaction. Benzoylation should block all three amino-acids. The advantage of this, and of the DNFB reaction (which becomes coloured only after the azo-dye is synthesized on to the protein-bound DNFB) is that they are stoichiometric reactions which can be measured quantitatively.

METHOD

The sections need not be fixed for this procedure. [Should fixation be required (e.g. to free more of the potentially reactive groups) then fix the cryostat sections for 10 min in acetic-alcohol (5 : 95 v/v or 1 : 3 v/v). Take the sections through graded concentrations of alcohol to water before proceeding to 1.]

IN ICE BATH (with pre-cooled solutions)

1. Place slide in freshly prepared cooled solution of tetra-azotized dianisidine for 6 min.
2. Transfer to another freshly prepared solution of tetra-azotized dianisidine for a further 6 min.

   Note: this solution should be prepared during the first 6 min reaction. The reason for fresh solutions is that the tetra-azotized dianisidine decomposes very rapidly so that one can be certain of good activity for only about 6 min.
3. Wash well in 2% cooled barbiturate solution.
4. Wash well in cooled distilled water.

   Note: it is essential to remove adsorbed, unreacted tetra-azotized dianisidine because it will also be intensified in step 5 and yield a strong, but spurious stain. The reacted compound is held by true chemical linkage and cannot be washed out of the protein.]
5. Stain for 15 min in cold H-acid solution in a Coplin jar.
6. Wash in cold distilled water.

Remove from ice bath.

7. Dehydrate and mount in Euparal or DePex.

Result. Red or brown-red.

Solutions required for this technique

Barbiturate solution

Sodium diethyl barbiturate (Veronal) 6 g

Make up to 300 ml with distilled water.

Tetra-azotized dianisidine solution

Grind 0·15 g tetra-azotized dianisidine with cooled barbiturate solution in a cooled mortar until dissolved. Make up to 100 ml with barbiturate solution.

H-acid

H-acid	0·5 g
Sodium bicarbonate	1·0 g

Make up to 100 ml with distilled water.

The bicarbonate is a convenient way of achieving an alkaline pH.

Pretreatment with performic acid

Solution

30% (100 vol) hydrogen peroxide	4 ml
Conc. sulphuric acid	0·5 ml
98% formic acid	40 ml

The performic acid should be vigorously stirred before use to remove gas and the peroxide must be fresh.

Place the section in this solution for 45 min at room temperature.

BAKER'S METHOD FOR **TYROSINE** (Baker, 1956)

Rationale. The phenolic amino-acid, tyrosine, is a useful marker for most proteins (but not for collagen). The most used histochemical procedure for this amino-acid has been a modification of the chemical Millon's test (e.g. Bensley and Gersh, 1933). Baker developed the following technique from the very sensitive biochemical Folin reaction (Folin and Marenzi, 1929) and it yields a fairly strong stain in 10 μ sections. In this reaction (Gibbs, 1927; Baker, 1956) the phenol is converted into a nitrosophenol by the sodium nitrite and the mercury links to the nitrogen atom of this nitroso group to form a red compound.

METHOD

Fresh cryostat sections may be fixed in acetic alcohol (5% glacial acetic in absolute alcohol) for 10 min and then taken through graded concentrations of alcohol to water.

1. Put 25 ml of mercuric sulphate solution into a 50 ml beaker.
2. Add 2·5 ml sodium nitrite solution.
3. Put the section in the beaker and heat gently until the mercuric sulphate–sodium nitrite solution is just boiling.
4. Remove the sections. Wash well in three changes of distilled water.
5. Either mount in glycerine or glycerine-jelly or dehydrate and mount in Euparal or in DePeX.

Result. Phenols, especially tyrosine and hence most proteins, stain red, pink or yellow-red.

Solutions required for this technique

Mercuric sulphate solution

Add 10 ml conc. sulphuric acid to 90 ml distilled water.

Then add 10 g mercuric sulphate (preferably M.A.R. grade).

Heat to dissolve. Cool. Make up to 200 ml with distilled water. This solution is stable.

Sodium nitrite solution

Sodium nitrite 0·25 g

Make up to 100 ml with distilled water.

FAST GREEN METHOD FOR **BASIC PROTEIN** (see Alfert and Geschwind, 1953)

Rationale. At pH 8 the main groups which are still sufficiently ionized to bind acidic dyes are the guanidine groups of arginine and the ϵ-amino groups of lysine (e.g. see Klotz, 1950). They are usually bound to other acidic groups in the tissue, particularly to the nucleic acids. Hence these have to be removed by treatment with hot trichloroacetic acid. This has the additional advantage of washing out acid soluble basic matter of low molecular weight, i.e. which are not part of the protein molecule.

The basic groups unmasked by the acid treatment tend to bind acidic ions present in buffer solutions and these will interfere with the ability of the basic groups to bind Fast green. Consequently Alfert and Geschwind suggested that the dye solution should be prepared in distilled water and brought to pH 8 by judicious addition of alkali. However we find that citrate buffer is a more convenient way of achieving the required pH and does not produce this competitive inhibition.

METHOD

Fresh cryostat sections should be fixed in acetic alcohol (5 : 95 v/v) and then taken through graded concentrations of alcohol to water. (Alfert and Geschwind suggested fixation in 10% neutral formalin.)

1. Immerse the sections for 15 min in 5% trichloroacetic acid which is at 95°C (e.g. in a Coplin jar standing in a beaker of boiling water).
2. Wash well in 3 changes (10 min each) of 70% alcohol and then in distilled water to remove all the trichloroacetic acid.
3. Stain in Fast green for 30 min at room temperature.
4. Wash briefly in distilled water.
5. Transfer the section directly to 95% alcohol.
6. Dehydrate and mount in Euparal or in DePeX.

Result. Basic protein stains green or green-blue. Histones should be relatively strongly coloured.

Solutions required for this technique

Trichloroacetic acid

5% in water

Fast green FCF

0·1% in citrate-phosphate buffer, pH 8·0

Buffer (McIlvaine's)

0·1M citric acid (0·48 g in 25 ml)	2·75 ml
0·2M disodium phosphate (7·1 g Na_2HPO_4.anhydrous in 250 ml)	97·25 ml

DE-AMINATION METHOD

Terminal—NH_2 and ε-amino groups are 'destroyed' by this procedure.

Mix equal volumes of 0·1 N hydrochloric acid and 0·1 N sodium nitrite (6·9 g in 1000 ml). Leave sections in this freshly prepared solution for 10 min at room temperature. Transfer to fresh solution for a further 10 min.

ASSESSMENT OF 'pK' VALUES (see Levine, 1939; Klotz, 1950)

Rationale. For basic or acidic groups to be capable of asserting their basic or acidic properties, it is necessary that they should exist in the ionized form. For example, the carboxyl group —COOH is not actually acidic unless it exists in the ionized (charged) form —COO^-. Similarly the amino group —NH_2 is not basic unless it is ionized to —$\overset{+}{N}H_3$. The ionization of acidic groups becomes progressively suppressed below their pK value (i.e. at more acid pH values) and that of basic groups above their pK value (pK may be defined as that pH value at which the group is 50% ionized). In histochemistry the pK value cannot be estimated accurately but its assessment is often valuable. In part this difficulty of achieving an accurate estimate is due to the fact that ions in the solution, especially those of buffers, may affect the ability of dyes to react with the charged groups. But another difficulty is that of ascribing the same precise pK values for groups when they occur in tissue as when they are studied in isolated chemical systems. For all that, the data may give a helpful approximation. In proteins the groups which can contribute to the acidic or basic properties of the molecule are as shown in the formulae below, where the pK value is that of isolated, purified amino-acids (Fieser and Fieser, 1953).

From these values, which are only approximate, it would seem that protein should not be able to bind basic (positively charged) dyes at pH values much below pH 4 (at which value half the ionization of the —COOH groups of glutamic acid may be suppressed) nor should it bind acidic dyes (negatively charged dyes) appreciably at or above pH 8 unless it contains much lysine, arginine or cysteine (or possibly tyrosine).

The same considerations apply to any acidic and basic groups in tissue sections, not only to those in protein. A rough guide to the acidity of a region of a cell or tissue can be obtained by staining with acidic and with basic dyes over a range of pH values. However, greater precision is obtained if dilute solutions (10^{-4}M*) of the dye in very dilute buffer solutions (e.g. McIlvaine's citrate-phosphate buffer diluted ten times) are used for long periods (e.g. for 36 hr). The sections are then washed thoroughly but briefly in distilled water. They are blotted dry and put straight into tertiary butanol, after which they may be mounted in Euparal or in DePeX.

The basic dyes can be toluidine blue or methylene blue; the acidic dyes can be fast green, although other workers have used Ponceau 2R or orange G (but both of these are 'levelling dyes' and this may complicate the effect—see Baker, 1958).

Some of the amino-acids, shown as the residues attached to the peptide chain, which can contribute to the basic and acidic properties of a protein, are shown on p. 46. The terminal —NH_2 (pK 7·5–8·5) and —COOH (pK 3) are not shown.

* It is difficult to talk of the molar concentration of dye-stuffs because they are impure mixtures. We follow Levine in using molarity as a guide to the weight of dye to be used.

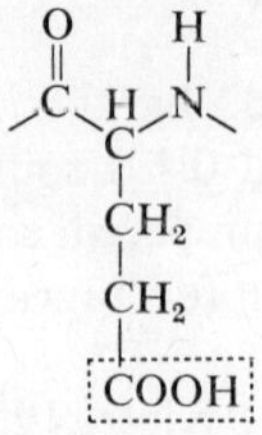

glutamate residue	histidine residue	tyrosine residue
The carboxyl group has a pK of 4	The imidazole group has a pK of 6–7	The phenolic group has a pK of 10

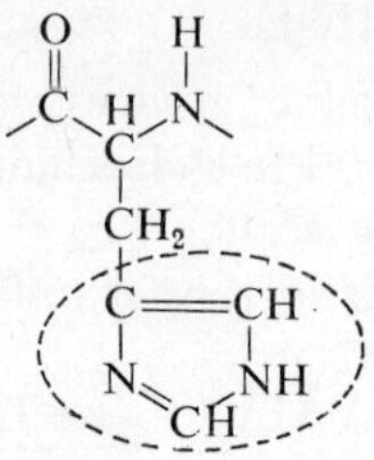

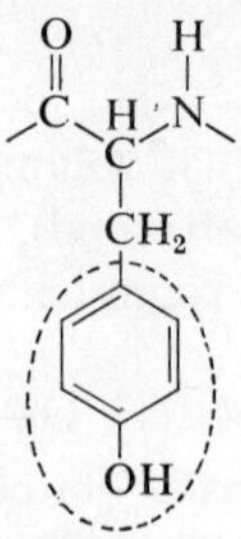

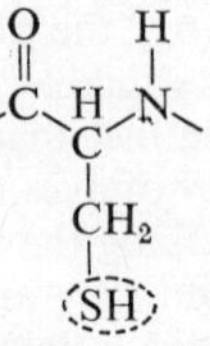

cysteine residue	lysine residue	arginine residue
The sulphydryl group has a pK of 9–11	The ε-amino group has a pK of 9–11	The guanidine grouping has a pK of 12–13

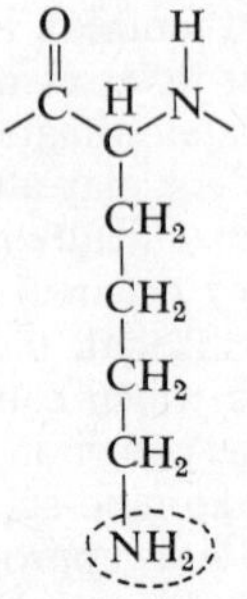

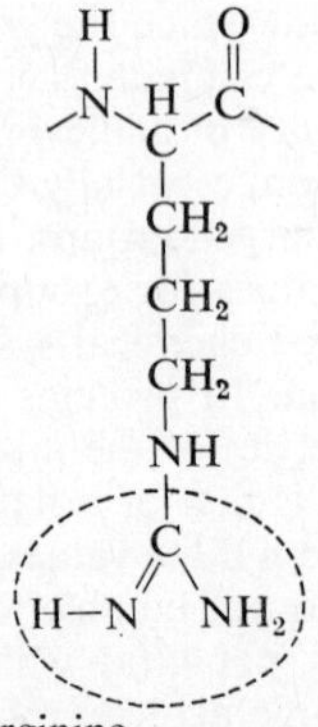

THE HISTOCHEMICAL SAKAGUCHI REACTION FOR ARGININE

Rationale. Biochemically the amino-acid is estimated by Sakaguchi's (1925) method in which α-naphthol is linked to the guanido-amino group of arginine at a very alkaline pH. Colour is then produced by the addition of hypobromite or hypochlorite. The specificity of the histochemical application of this reaction was investigated fully by Baker (1947). McLeish *et al.* (1957) used the dichloro-naphthol because this gives a stronger colour and reacts more rapidly, and used hypochlorite because it is easy to obtain. There are two hazards to the method: first, that after producing the colour, the hypochlorite must be removed because its prolonged action will reduce the final colour; and secondly that the mountant should be alkaline because the colour formed is a pH indicator, being straw coloured at low pH values but strongly red-brown at higher pH values. The final method (McLeish *et al.*, 1957; see also McLeish, 1959) can be used for the quantitative measurement of the amount of arginine available to be stained in the section. Sometimes it may be necessary to use special fixation to unmask and yet retain the arginine in tissue sections. This is the reason why La Cour *et al.* (1958) and Howe (1959) used Lewitsky's fluid; McLeish (1959) found formalin sufficient for his plant tissues.

METHOD

Cryostat sections should be fixed either in Lewitsky's fluid (15 min) or in 10% neutral formalin (McLeish and Sherratt, 1958; McLeish, 1959). (For this reaction the tissue itself may be fixed in either of these fluids and may be embedded in paraffin wax before it is sectioned.) Then proceed as follows:

1. Pass the section up the alcohol series to absolute alcohol.
2. Dip the whole slide into celloidin solution. Shake off excess; drain and air-dry.
3. Take the section through graded concentrations of alcohol to water.
4. Treat it in freshly prepared reaction medium for 6 min.
5. Rinse rapidly in 5% solution of urea (i.e. to stop the reaction. But note that prolonged treatment with urea will decolourize the reaction product.)
6. Place in 1% solution of sodium hydroxide for 5 min.
7. Mount in alkaline glycerol [the alkalinity is needed to retain the colour].

Result. Orange-red colour denotes arginine.

Solutions required for this method

Lewitsky's fluid

10% formalin (i.e. 10% of the 40% formaldehyde)	1 volume
1% chromic acid (chromium trioxide)	1 volume

Mix just before use.

It is advisable to wash well after this fixative (e.g. 5 min in each of three baths of distilled water).

Celloidin solution

Ether (diethyl ether)	50 ml
Absolute ethyl alcohol	50 ml
Celloidin	1 g

This is used to protect the tissue against the subsequent treatments, which tend to remove the section from the slide.

Reaction medium

1% aqueous solution of sodium hydroxide	30 ml
1% solution of 2,4-dichloro-α-naphthol in 70% ethyl alcohol	0·6 ml
'Milton' (proprietary brand of sodium hypochlorite)	1·2 ml

Mix immediately before it is to be applied to the sections.

[*Note:* the 'Milton' should be from a freshly opened bottle as its efficiency for this reaction becomes reduced after a week or two of having been opened.]

Urea solution

5%, aqueous

Sodium hydroxide solution

1%, aqueous

Alkaline glycerol

Glycerol	9 ml
10% aqueous solution of sodium hydroxide	1 ml

[*Note:* In their description of a histochemical Sakaguchi method, Carver *et al.* (1953) recommended dehydration into tertiary butyl alcohol; then into aniline oil for 3 min; wash in xylene for 10 sec; and make the preparation permanent by mounting in the following mountant:

Aniline	0·1 ml
Xylene	100 ml

Add one volume of this to four volumes of Permount.]

CHÈVREMONT–FRÉDERICK METHOD FOR —**SH GROUPS** (e.g. Chèvremont and Fréderick, 1943)

Rationale. The principle of this method is that sulphydryl groups are such strong reducing agents that they can reduce ferricyanide to ferrocyanide. The latter reacts with ferric ions to produce an intense and insoluble precipitate of Prussian blue, as in the normal chemical test for iron:

$$2[Fe(CN)_6]^{3-} + 2Pr{-}SH \rightarrow 2[Fe(CN)_6]^{4-} + Pr{-}S{-}S{-}Pr$$

ferricyanide ions + protein-bound sulphydryl (e.g. cysteine) → ferrocyanide + cross-linkage between protein-bound sulphur atoms (e.g. cystine)

$$3K_4[Fe(CN)_6] + 4Fe^{3+} \rightarrow Fe_4[Fe(CN)_6]_3 + 12K^+$$

potassium ferrocyanide + ferric ions → Prussian blue (ferric ferrocyanide)

It is true that this reaction will be given by any strongly reducing substance (as discussed by Gomori, 1952) and for this reason it has fallen into some disrepute. In practice, all this means is that it requires to be controlled by testing a serial section which has had the sulphydryl groups blocked specifically. The sites which are negative in the control but positive in the test contain sulphydryl groups.

Frequently confusion has occurred because not all the sulphydryl groups have been made available to the reaction. For example, many —SH (sulphydryl) groups may occur as the fully oxidized —S—S— (disulphide) group, and these will require to be reduced to —SH before they can react. The advantage of this ferricyanide method is its intensity, and its reliability (in our experience, at least). [*Note:* other methods, which seem to be potentially valuable, depend on the reduction of tetrazolium salts—as suggested by Barrnett and Seligman, 1952, and Gomori, 1956, or on the use of tetra-azotized dianisidine as on p. 41 and as recommended by Gomori, 1956, and by Casselman, 1959.]

METHOD

Unfixed cryostat sections can be used directly. Otherwise the sections may be fixed for 15 min in trichloroacetic-ethanol (1% trichloroacetic acid in absolute or 80% alcohol).

1. Immerse the sections in three changes of the ferricyanide solution giving a total reaction time of 20 min.

(Paraffin-embedded sections may require a total reaction time of 25 min or longer.)

[*Note:* the ferricyanide solution should be prepared just before it is to be used.]

2. Wash in distilled water.

3. Either mount in Farrants' medium or dehydrate and mount in Euparal or in DePeX.

Result. SH and other strong reducing sites stain blue.

Control. The specificity of this method depends on the use of an adequate and specific blockade of —SH groups. Either of the following blockades can be used; both have been useful in our experience:

Blockade 1. Treat sections for 24–48 hr in a saturated aqueous solution of mercuric chloride. (Excess mercury may be removed by washing in Lugol's iodine, followed by treatment with a 5% solution of sodium thiosulphate, provided that the mercury which is blocking the —SH groups is not affected.) Then test with the ferricyanide method: —SH groups should not react after this blockade.

Blockade 2. Immerse sections for 2–3 days in a saturated solution of phenylmercuric chloride in butyl alcohol.

[*Note:* Casselman, 1959, suggested that 0·1M N-ethyl maleimide in phosphate buffer at pH 7·4 for 4 hr at 37°C, or the use of 0·1M iodoacetic acid, brought to pH 8·0, at 37°C for 20 hr. are very effective blocking agents for sulphydryl groups.]

The Demonstration of —S—S— Groups

Many substances contain an equilibrium mixture of —SH and —S—S— forms. Others, particularly the adult form of keratin, contain predominantly the —S—S— form. This has to be reduced to —SH before any of the sulphydryl reactions can give a positive result. Casselman (1959) recommended the use of either thioglycollic acid (Barrnett and Seligman, 1952, 1954) or of thioglycerol (Gomori, 1956). We have immersed sections in a 20% aqueous solution of sodium dithionite (as for the DNFB method) at 37°C for 30 min although this requires to be well washed to remove excess dithionite. Alternatively we have found that a solution of ascorbic acid in water gives comparable results and is more physiological.

Sodium sulphite or sodium cyanide (5% aqueous solutions used for 10 min) have also been recommended for converting —S—S— to —SH (Serra, 1946; Gomori, 1952).

TO DEMONSTRATE —S—S— GROUPS

Take three serial sections.

1. Block pre-existing —SH groups, e.g. by treatment with mercuric chloride.

2. Remove free mercury by rinsing in Lugol's iodine and then in thiosulphate solution.

3. Rinse in water.

4. Treat one of these sections with the ferricyanide method for sulphydryl groups, or with Bennett's method. Any positive staining should be due to non-specified reducing sites.

5. Treat the remaining two slides with a reducing solution such as warm dithionite. The exact time needed for this reduction may vary with the tissue but 30 min is suggested. This should convert —S—S— to —SH.

6. Treat one of these slides with a sulphydryl-blocking method (such as mercuric chloride).

7. Stain both the slides with, for example, the ferricyanide method. The unblocked section should be blue wherever disulphide bonds had occurred; the proof that these were disulphides (but now have been converted to sulphydryl) will be that the blocked section does not show a blue colour at these sites.

Since pre-existing sulphydryl groups had been blocked (step No. 1 in this sequence), the specific blue colour will be due solely to —S—S— bonds which have been converted to —SH by whatever procedure is used in step No. 5. Should it be wished merely to demonstrate all sulphur-containing amino-acids, irrespective of whether they occurred originally in the oxidized or in the reduced state, then steps Nos. 1, 2 and 3 can be omitted. The result obtained from step No. 4 will then represent free, pre-existing —SH (provided that no colour is produced if a suitable blockade is used) and the results from step No. 7 will represent total sulphur-containing amino-acids (although as with the other results, the specificity will depend on the lack of colour at the sites after suitable blockade).

Solutions required for these methods

Ferricyanide solution (containing ferric ions)

Solution 1:	Ferric ammonium sulphate	2·4 g
	$(NH_4)_2SO_4Fe_2(SO_4)_3.24H_2O$	
	Distilled water to	100 ml

This can be kept as a stock solution and merely provides the free ferric ions.

Solution 2:	Potassium ferricyanide	0·1 g
	Distilled water to	100 ml
	Prepare this solution just before use.	

The reaction medium is prepared by adding 3 volumes of Solution 1 to 1 volume of Solution 2 just before it is to be used. Then adjust the pH to pH 2·4.

Blockade. Either a saturated solution of mercuric chloride in water (by 'saturated' is meant a solution which also contains solid mercuric chloride; the solution is left to settle and the supernatant fluid is used); or a saturated solution of phenylmercuric chloride in butyl alcohol. The advantage of the latter may be that it may minimize the presence of globules of free mercury in the tissue section.

Lugol's iodine

Iodine	1 g
Potassium iodide	2 g
Distilled water to	100 ml

Thiosulphate solution

5% aqueous solution of sodium thiosulphate.

BENNETT'S MERCURIAL METHOD FOR SULPHYDRYL GROUPS
(see Bennett and Watts, 1958)

Rationale. Mercurials react specifically with sulphydryl groups. This is the basis of the specific blockade methods (including the use of chloromercuribenzoate) discussed under the Chèvremont and Fréderick ferricyanide method (above, p. 49). Consequently if the mercury were to be attached to a dye, and yet be available to 'block' the sylphydryl groups, then the presence of the dye would mark the location of these groups. Hence Bennett developed 'mercury orange', namely 1-(4-chloromercuriphenylazo)-2-naphthol, for the demonstration of —SH groups:

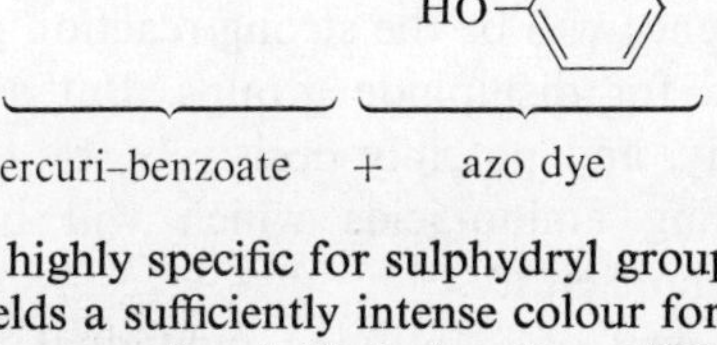

Cl—Hg—⟨C₆H₄⟩—N=N—⟨naphthyl⟩—OH

i.e. chloro–mercuri–benzoate + azo dye

This reagent should be highly specific for sulphydryl groups. There is some doubt as to whether it yields a sufficiently intense colour for most qualitative work. The same problems, namely of disclosing the —SH groups (e.g. trichloroacetic acid at room temperature may unmask some) and of converting disulphides to the sulphydryl form before a reaction can be obtained, apply to this method as they do to the Chèvremont and Fréderick technique (where they are discussed more fully—see p. 49).

METHOD

For fixation, a 1% solution of trichloroacetic acid in 80% ethanol is recommended by Barka and Anderson (1963). If cryostat sections are used they have to be taken to absolute alcohol before proceeding as follows:

1. Immerse for 16–24 hr in mercury orange solution at room temperature. The duration of the reaction does not seem to be critical.
2. Wash in two changes of absolute ethyl alcohol.
3. Mount in Euparal.

Result. Orange-red colour denotes sulphydryl groups. Most workers have reported rather weak reactions.

Blockade control. As for the ferricyanide method (above).

Solutions required for this method

Mercury orange solution

Bennett and Watts recommended 1 mg of mercury orange in 100 ml of toluene. Others have suggested that a saturated solution of mercury orange in absolute alcohol (about 3 mg in 100 ml) should be prepared and then diluted to make the final concentration of the alcohol 80%. Some have suggested using an 80% ethanolic solution buffered to pH 8 (see Barka and Anderson, 1963, pp. 32–34).

IMMUNOFLUORESCENCE METHODS FOR PROTEINS

The methods described so far do not define the actual protein molecule but rather the active or available side-groups of the protein. For example, collagen lacks tyrosine and hence although it stains well with the dinitrofluorobenzene method, it is negative to Baker's tyrosine technique. Equally histone proteins have relatively high concentrations of basic amino-acids such as lysine and arginine, and these can be demonstrated by methods described in the preceding pages. Perhaps more striking in general histochemical experience will be the strong reaction given by keratin to the ferricyanide method for disulphide groups. But although the keratin will be stained distinctly, and possibly decisively, by this technique, it will be the sulphur-containing amino-acids which will be stained, not the keratin molecule as a whole entity.

The rational way of identifying individual tissue proteins as molecular entities is by the immunological method developed by Coons (see Coons,

1952; 1956; 1958). The essence of this technique is that each protein may act as an antigen, and so produce a highly specific antibody to itself, if it is injected into a suitable animal. The antibody, which is then found in the globulin fraction of the injected animal's blood, should form a precipitate with the antigen, but with no other protein, because the precipitation reaction appears to be of a specific 'lock-and-key' type (e.g. see Haurowitz, 1950).

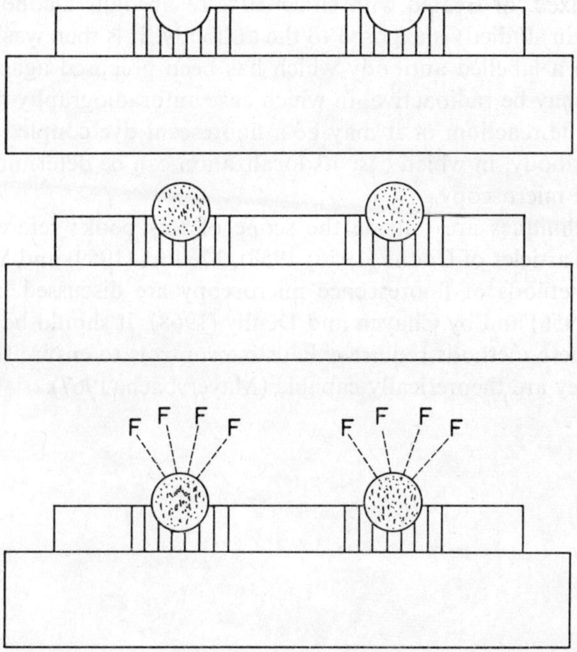

Fig. 13. The antigen in the section (hollowed out hatched areas) reacts specifically with the antibody (rounded globulin molecules) which in turn links the fluorescein-labelled antiglobulin (marked F). The point of this sandwich technique is to produce many fluorescent molecules at the site of each antigenic group, without impairing the highly specific reaction between the specific antigenic site and its antibody. The antiglobulin reaction is less delicate and specific.

The antibody can retain its specificity for this precipitation reaction even if it has been conjugated with fluorescent dyes (Marrack, 1934; Coons *et al.*, 1942), so that the site of the reaction can be rendered apparent by the fluorescence of the dye which is linked to the antibody (which has reacted specifically with the antigen). Moreover, since the antibody will be a globulin, the intensity of the reaction can be greatly accentuated if the antibody is allowed to react with the antigen and then a labelled antibody to globulin (which need be specific only for this general class of proteins and not for that particular antibody) be allowed to react with the precipitate and localize it. (This is the 'sandwich technique', Fig. 13.)

In practice this procedure requires the isolation of the protein whose histochemical localization is being sought. An antibody has to be prepared against this protein, or this species of protein. This is not quite as impracticable as it sounds because antibodies to various types of protein have been made by various workers and some are available commercially. The tissue section (unfixed, or treated with either 80% or absolute alcohol, depending on the protein studied) is exposed to the antibody. It is then washed well and treated with a labelled antibody which has been prepared against globulin. The 'label' may be radioactive, in which case autoradiography must be used to localize the reaction, or it may be a fluorescent dye coupled to this anti-globulin antibody, in which case its localization can be determined by direct fluorescence microscopy.

These techniques are outside the scope of this book; reference may be made to the articles of Coons (1956; 1958), Mellors (1959) and White (1960), while the methods of fluorescence microscopy are discussed by Price and Schwartz (1956) and by Chayen and Denby (1968). It should be noted, however, that these methods require exhaustive controls to ensure the specificity of which they are theoretically capable (Mayersbach, 1967).

STAGE 3: REACTIONS FOR ACIDIC COMPOUNDS, e.g. THE NUCLEIC ACIDS

THE FEULGEN REACTION FOR **DEOXYRIBONUCLEIC ACID (DNA)**

Rationale. Neither DNA nor ribonucleic acid (RNA) contain free aldehyde groups. However acid hydrolysis can unmask aldehyde groups in the deoxyribose moiety of DNA, but not in the ribose moiety of RNA. The precise chemical reason for this is not clear; it may be related (see Brown and Todd, 1955) to the ability of the phosphate to migrate to different sites in the sugars (so that the lack of one oxygen atom in the deoxyribose moiety is critical) and so allow the ring form of the sugar to be opened up to give a straight-chain molecule.

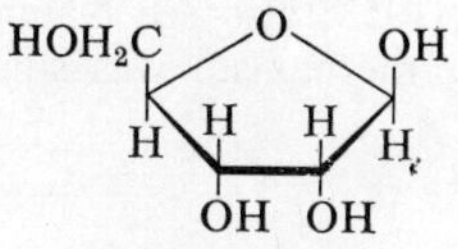

HOH$_2$C, O, OH; H, H, H, H; OH, OH

D-Ribofuranose
The ring form of ribose
as found in RNA

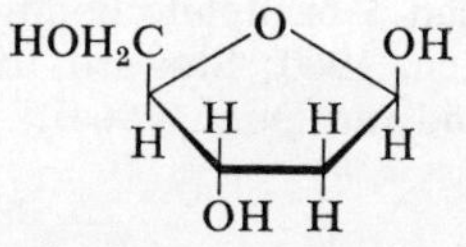

HOH$_2$C, O, OH; H, H, H, H; OH, H

D-2-Deoxyribofuranose
The ring form of deoxyribose
as found in DNA

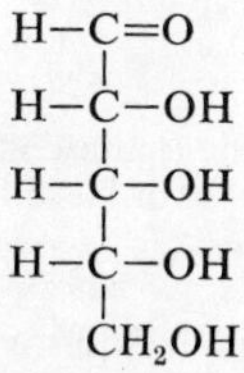

H—C=O
|
H—C—OH
|
H—C—OH
|
H—C—OH
|
CH$_2$OH

Straight-chain form of
ribose

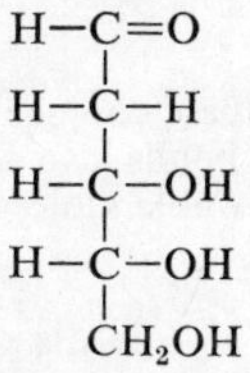

H—C=O
|
H—C—H
|
H—C—OH
|
H—C—OH
|
CH$_2$OH

Straight-chain form of
deoxyribose

It will be seen that in the straight-chain form, both ribose and deoxyribose have a terminal aldehyde (H—C=O) group.

Another chemical difference between ribose and deoxyribose is the fact that only the latter can form levulinic acid when treated with dilute acid:

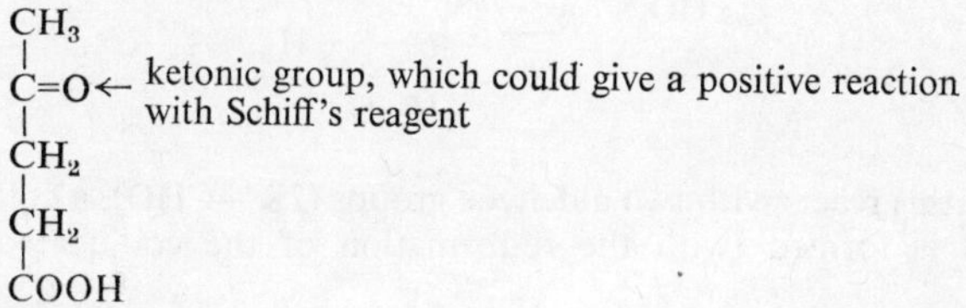

CH$_3$
|
C=O ← ketonic group, which could give a positive reaction with Schiff's reagent
|
CH$_2$
|
CH$_2$
|
COOH

Although the chemistry is not properly understood (see Davidson, 1960) the Feulgen nucleal reaction (to distinguish it from the Feulgen plasmal reaction) is as follows:

1. There is no aldehyde (or ketone) group at the beginning of the method.

2. Mild acid hydrolysis unmasks aldehyde groups in DNA but not in RNA. During this process (1 N hydrochloric acid at 60°C for a few minutes) the RNA is, in fact, removed from the tissue (but it still does not colour with Schiff's reagent); the acid also removes purines from the DNA. This release of purines is essential for the unmasking of the aldehyde group in the deoxyribose moiety. A balance must be achieved: on the one hand, hydrolysis removes the purines from the DNA and so gives more aldehyde groups for staining; on the other hand, prolonged hydrolysis splits the inter-nucleotide bonds and so renders the DNA soluble so that it becomes progressively lost from the section. Fixation affects the binding of the DNA to the nuclear proteins and so can retard both the loss of purines and the susceptibility of the inter-nucleotide bonds to the hydrolysis. (Consequently different times of hydrolysis are required after fixation in different fluids.)

3. The aldehyde groups are then exposed to Schiff's reagent which links to them through a bisulphite bond (see Lessler, 1953; also see Nauman *et al.*, 1960; Kasten, 1960; Stoward, 1966; another mechanism was suggested by Hardonk and van Duijn, 1964):

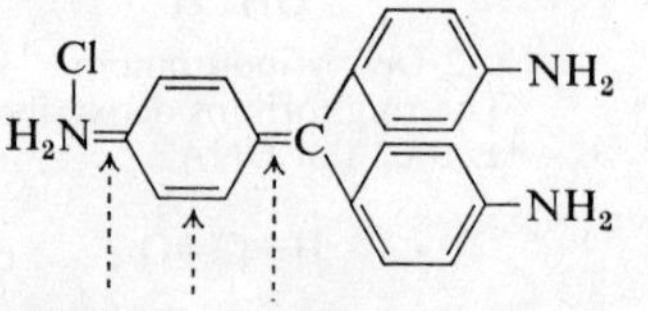

A

Note: these conjugated double bonds give colour to this whole molecule

A. Pararosaniline (main constituent of basic fuchsin)

When this dye is dissolved in sulphurous acid it is converted to the leuco form (colourless because the conjugated double bonds have been lost) and it forms a complex with the sulphurous acid:

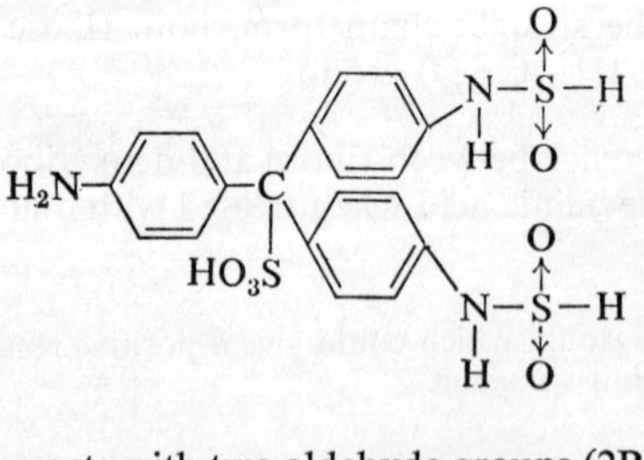

B

When this reacts with two aldehyde groups (2R—CHO), a coloured addition complex is formed (with the re-formation of the conjugated double-bond system).

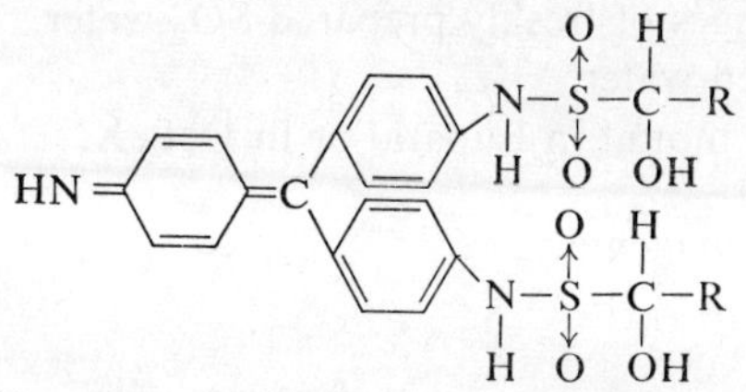

```
                              O  H
                              ↑  |
                           N—S—C—R
                           |  ↓  |
                           H  O  OH
HN=⟨ ⟩=C⟨
                              O  H
                              ↑  |
                           N—S—C—R
                           |  ↓  |
                           H  O  OH
```

4. After treatment with Schiff's reagent the unbound reagent must be washed out of the section. If this is not done then it will oxidize very readily, i.e. the leuco-form (B) will be converted back into the pararosaniline form (A). This is not a Schiff-complex and is spurious. [*Note:* for this reason it is nonsensical to wash the sections with tap water or with water to which an oxidant has been added because these will produce the coloured pararosaniline form from any adsorbed leuco-basic fuchsin. Yet some workers actually recommend that this be done. It will certainly increase the amount of colour in the section but no one will be able to assess how much of this colour is a true Feulgen reaction, and how much is due to adsorbed re-oxidized fuchsin.]

Consequently the section is washed in SO_2-water. This will couple any free pararosaniline (A) to produce the leuco-form (B) and will wash it out of the section.

5. For quantitative work particularly it should be noted that the Schiff's reagent loses SO_2 on being kept for any length of time, and the pH of the reagent becomes altered. The intensity of the stain (and its spectroscopic characteristics) depends fairly critically on the pH at which it is used.

Precautions. The Feulgen reaction depends on the production of aldehyde groups (which react with the Schiff's reagent to produce a magenta dye) as a consequence of hydrolysis in 1 N hydrochloric acid at 60°C. The presence of a colour (when the reaction is performed correctly) shows that such groups have become available during this procedure: it is not yet proof that they required the hot acid for their unmasking. Other potential aldehydes can be unmasked just by cold acid (e.g. the plasmals are characterized in this way). Hence it is necessary to use one serial section as a control: it is placed in 1 N hydrochloric acid at room temperature for 15 min and then exposed to the Schiff's reagent. It is the difference between the test and the control section which demonstrates the presence of DNA. [*Note:* in general such a control is negative or only weakly coloured relative to the test section so that such a control need not be done invariably.]

METHOD

1. Fix cryostat sections in acetic-ethanol (1 : 3 v/v) for 10 min.
2. Take through graded concentrations of alcohol to water.
3. Rinse in 1 N hydrochloric acid.
4. Plunge into 1 N hydrochloric acid which is at 60°C. Maintain at this temperature for 6 min.
5. Transfer to 1 N hydrochloric acid at room temperature.
6. Immerse in Schiff's reagent in the dark for 60 min.

7. Rinse in 3 changes of freshly prepared SO_2-water.
8. Rinse in distilled water.
9. Dehydrate and mount in Euparal or in DePeX.

Result. DNA stains magenta.

Control

1. Fix serial section in 1 : 3 acetic–ethanol for 10 min.
2. Take down to water.
3. Rinse in 1 N hydrochloric acid and leave in this at room temperature for 15 min.
4. Immerse in Schiff's reagent in the dark for 60 min.
5. Rinse in three changes of freshly prepared SO_2-water.
6. Rinse in distilled water.
7. Dehydrate and mount in Euparal or in DePeX.

Any stain produced by this control procedure is *not* due to DNA and must be subtracted from the results obtained by the full test procedure.

Additional control. Deoxyribonuclease can be used to make the results more rigorous (see later).

Solutions required for this procedure

Schiff's reagent

Dissolve 1 g basic fuchsin in 200 ml of boiling distilled water. Shake and cool to 50°C. Filter and add 30 ml of 1 N hydrochloric acid and 3 g of potassium metabisulphite. This solution is allowed to stand for 24 hr in a dark-coloured, stoppered bottle. During this time the bisulphite decolorizes the fuchsin, i.e. it forms the colourless SO_2-pararosaniline complex (leuco-basic fuchsin). Impurities and other coloured matter are then removed by adsorption on to activated charcoal. Therefore, add 0·5 g of a decolorizing activated charcoal ['Norit', a commercial vegetable charcoal has been recommended—the properties of the charcoal used may be critical], and shake well for about 1 min. Filter rapidly (to avoid re-colorizing the leuco-basic fuchsin) through a coarse filter paper (or glass-wool). The solution should be clear and colourless; it can be stored for months in a well-stoppered, dark-coloured bottle.

This description of how to prepare Schiff's reagent is given primarily to demonstrate its constitution. It is often advisable to buy commercially prepared Schiff's reagent from a suitable source (see Appendix).

SO_2-water

1 N hydrochloric acid	25 ml
0·5% aqueous solution of potassium metabisulphite	50 ml

Mix immediately before use.

1 N *hydrochloric acid*

Analar hydrochloric acid sp. gr. 1·18	89 ml
Distilled water	911 ml

Always add the acid to the water.

KURNICK'S METHYL GREEN METHOD FOR DNA

Rationale. Kurnick (1949, 1950; Kurnick and Mirsky, 1949) showed that triphenyl methane dyes which contain two charged amino groups stain polymerized nucleic acids selectively; if they have only one charged group (e.g. crystal violet) they do not show this selectivity:

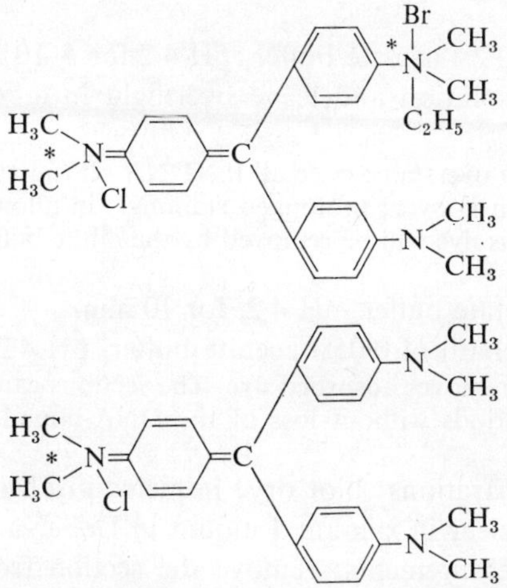

methyl green

crystal violet

Both Kurnick and also Vercauteren (1950) suggested that the specificity shown by methyl green for polymerized DNA was due to the actual distance between these basic groups: they were spaced just sufficiently to fit on two phosphate moieties of the DNA, one above the other in the double helix.

Whether these ideas be correct or not, it has been shown that methyl green can be made to bind to DNA so strongly that it is not washed out even by dilute buffer; moreover this binding is stoichiometric, so that one molecule of methyl green bound in the section is equivalent to a length of the DNA corresponding to 20 phosphate groups in the DNA molecule (see Kurnick, 1950).

One technical problem must be stressed. The phosphate groups of the DNA are not freely available in cells; they are usually linked to basic groups, such as the arginine and lysine moieties of nuclear histones (e.g. see Kurnick and Mirsky, 1949). Consequently some acidic treatment may be required to render the phosphate groups available to the dye.

METHOD

Fresh cryostat sections can be used; but it must be realized that they will become fixed when they are immersed in the acetate buffer at pH 4·2 or by the pretreatment. Consequently it may be advisable to fix the fresh sections deliberately (rather than by the first coagulating fluid to which they are subjected). Absolute or 80% alcohol can be used for 10 min. Should this cause the chromatin to become too coagulated (and so unsuitable for quantitative measurement), then neutral formalin can be used (e.g. for 15 min).

For optically homogeneous nuclei, the sections should be treated with 10% sucrose for 15–30 min before they are fixed either in neutral formalin or in formol-sucrose (Ris and Mirsky, 1949).

1. Immerse sections in 0·1 N hydrochloric acid for 5 min (to split the DNA from histone-like protein).

2. Rinse in water.

3. Place the sections in 0·2M acetate buffer, pH 4·2 for 5–10 min.

4. Transfer to the dye solution and leave overnight in a refrigerator at about +4°C.

[*Note:* it is impossible to overstain: once all the DNA phosphate groups have been stained the reaction is over; prolonged staining will allow more dye to become adsorbed but this dye will be removed by the dilute buffer used in the next step.]

5. Immerse in 0·05M acetate buffer, pH 4·2, for 10 min.

6. Transfer to a second bath of 0·05M acetate buffer, pH 4·2, for 10 min.

[*Note:* the dilute buffer removes adsorbed dye. The sections can be left in this buffer for prolonged periods without loss of the DNA-bound dye (Chayen, 1952).]

7a. For permanent preparations: blot dry; immerse in absolute alcohol, and mount in Euparal or clear in xylol and mount in DePeX.

7b. For quantitative measurements: remove the section from the 0·05M acetate buffer and mount it direct in buffered glycerine.

Result. DNA stains green or green-blue.

[*Note:* some other polymeric molecules may also stain by this method. Also it should be noted that DNA which is not highly polymerized (either because of disease processes, or technical artifact) will not stain by this procedure.]

Control. The use of deoxyribonuclease (see below).

Solutions required for this method

Buffers. Please see Appendix.

Dye solution

[*Note:* methyl green very readily becomes degraded to methyl violet which interferes with the specific reaction for DNA. It must be removed before the dye solution is used.]

Dissolve 0·25 g of methyl green (a good brand, with relatively little impurity is strongly recommended; e.g. the Revector brand of Hopkin and

Williams has usually given good results) in 100 ml of 0·2M acetate buffer at pH 4·2.

Shake vigorously for several hours.

Put this solution into a large separating funnel.

Add 100–200 ml of chloroform. Shake well. Leave to settle and remove the chloroform (which will be full of methyl violet). Add another sample (100–200 ml) of chloroform. Shake; allow to settle; remove the chloroform.

Repeat until the chloroform is almost colourless.

Leave the aqueous dye solution in an open vessel in the dark to allow entrapped chloroform to evaporate.

This solution can be used immediately; it can be stored in a stoppered, dark-coloured bottle but it should be tested by shaking with chloroform each time before it is used, to ensure that there has not been excessive degradation to form methyl violet.

Buffered glycerine

Glycerine	4 volumes
0·2M acetate buffer at pH 4·2	1 volume

THE USE OF DEOXYRIBONUCLEASE

Rationale. The final proof that a substance in sections is indeed DNA must be that it is removed specifically by deoxyribonuclease. But even the 'crystalline' enzyme may contain proteolytic activity; the removal of a protein, to which may be attached an acidic substance, could simulate removal of DNA. To inhibit such proteolytic activity, 0·05M cysteine or hydroxylamine may be added to the incubation medium (see Jacobson and Webb, 1952). Magnesium ions may be included to activate the deoxyribonuclease even though most tissues may contain enough of this element.

As a control for the specific action of the deoxyribonuclease, a serial section can be treated with the same incubation medium, for the same time, but without the addition of deoxyribonuclease to the solution. A more rigorous control is to use the same incubation medium, containing deoxyribonuclease, but also containing zinc to inhibit the action of the deoxyribonuclease. This ensures the presence of the enzyme, but in an inactive form.

INCUBATION SOLUTION

The final concentration of the deoxyribonuclease is to be 0·1%. First prepare a 0·05M solution of cysteine hydrochloride (or hydroxylamine hydrochloride) in 0·05M sodium acetate solution at pH 7. Dissolve enough of the deoxyribonuclease in this so that the final concentration of the enzyme will be 0·1% (e.g. dissolve 2 mg of the deoxyribonuclease in 1 ml of the cysteine-acetate buffer solution).

Adjust to pH 7 if this should be necessary. Leave the solution to stand at room temperature (or at 37°C) for 5–15 min. (This is to inactivate the proteolytic activity.)

Before it is to be used, dilute this solution with an equal volume of barbitone-HCl buffer at pH 7.

To the enzyme solution add 1% of magnesium chloride.

Treatment. Add the complete medium to the section and leave at room temperature. One hour should be sufficient to remove DNA from unfixed sections. Then stain with methyl green as above.

To confirm the specificity of the deoxyribonuclease. Should it be necessary to prove the specificity of the deoxyribonuclease used, the following procedure should be followed:

Take three serial sections.

Treat one with the deoxyribonuclease incubation medium for 1 hr.

Treat the second for the same time with an exactly similar incubation medium to which has been added zinc sulphate (at a concentration of 0·1–0·01M; 0·1M sodium arsenate may be used instead of zinc sulphate). This should inhibit the deoxyribonuclease.

Leave the third section untreated (or it may be treated with the incubation medium lacking the enzyme entirely). Then stain all three slides together by Kurnick's methyl green method.

Result. DNA should stain green in the third slide; it should not be present in the first slide (having been removed by the active deoxyribonuclease). It should be present and stained green in the second slide (which has been subjected to the inactivated deoxyribonuclease).

THE METHYL GREEN-PYRONIN METHOD FOR DNA and RNA

Rationale. As has been discussed in connection with Kurnick's methyl green procedure, it is possible to obtain specific and quantitative staining of DNA by prolonged reaction with methyl green under controlled conditions. This specificity is said to be due to the two charged groups on the dye molecule both becoming linked to phosphate moieties in the DNA. The use of the mixture of basic dyes, methyl green and pyronin, to stain nuclei and cytoplasm differentially, is of long standing; however the mechanism by which this mixture causes this differentiation has been understood only relatively recently (Chayen, 1952; Brachet, 1954). It seems that under the conditions of this reaction, and in particular with the relatively short periods of staining used, the methyl green cannot achieve its completely stable binding to the DNA. Thus it seems that the mechanism of this reaction is the competition between the slow-staining, but doubly charged methyl green and the more rapidly staining, singly charged pyronin. The pyronin drives the methyl green out of all its sites of linkage except where its double charge gives it some selective advantage, namely when it responds to an acidic polymer such as DNA. Consequently the methyl green stains DNA, and retains its binding to this substance against the competitive action of the pyronin. On the other hand, the pyronin can stain the less polymerized ribonucleic acid (RNA) more rapidly, and can displace methyl green from its linkage with smaller polymeric acidic substances (such as RNA—but this applies also to acidic polysaccharides and similar molecules).

The reaction is done at pH 4·2 at what is about the iso-electric point of isolated nucleic acids (Kurnick and Mirsky, 1949). This has the advantages of having them in the charged condition and also in their least soluble state.

METHOD

Fresh cryostat sections can be used but it must be appreciated that they will be fixed by the acidic acetate buffer used in the staining mixture. Consequently the sections can be fixed in acetic-alcohol which is a good precipitant of nucleoprotein. The higher the concentration of the acetic acid that is used, the more will the nucleic acids be freed from the protein and hence the stronger the stain. Consequently 1 : 3 (v/v) acetic-alcohol may be used, although 5% acetic acid in absolute ethyl alcohol is satisfactory. Fixation should be for 15 min.

1. If the sections have been fixed in an alcoholic fixative, rinse in 70% alcohol and transfer to acetate buffer.
2. Immerse in stain mixture for 30 min.
3. Rinse very briefly in acetate buffer.
4. Blot dry. This must be done carefully but firmly. (If any moisture is left, the methyl green will be lost from the sections at the next step.)
5. Immerse in absolute *n*-butyl alcohol for 2 min.
6. Mount in Euparal.

Solutions required

Acetate buffer at pH 4·2

A. 0·2M sodium acetate solution consists of 6·8 g $CH_2.COONa.3H_2O$ made up to 250 ml in distilled water.

B. 0·2M acetic acid contains 2·85 ml of glacial acetic acid made up to 250 ml with distilled water.

Add 70 ml of B to 30 ml of A.

Stain mixture. First prepare a 0·25% (w/v) solution of methyl green in 0·2M acetate buffer at pH 4·2. Extract with chloroform (as for Kurnick's methyl green method) until almost all the methyl violet impurity has been removed in the chloroform. Separate the aqueous methyl green solution from the chloroform and leave to stand, uncovered, in the dark until the dissolved chloroform has evaporated.

Then add an equal volume of an 0·5% (w/v) solution of pyronin G in 0·2M acetate buffer. Mix. This solution is stable for some weeks.

[*Note:* an alternative method for plant histochemistry is to use the phenolic mixture recommended by Darlington and La Cour, 1947.]

Result. Nuclei stain green or blue; nucleoli and cytoplasm stain red.

[*Note:* mast cells stain a characteristic flame orange-red by this method. Plasma cells stain so intensely red that this procedure has been used to identify them.]

Interpretation. Very often, as a first approximation, the red colour is said to denote the presence of RNA and the blue or green colour that of DNA. But it must be emphasized that this can be used only as a rough approximation. In practice, when nuclei stain blue-green, this is probably a true indication of the presence of DNA; however, should the presence of DNA be in the least critical, this apparent localization should be confirmed by the use of deoxyribonuclease as discussed on p. 61. Alternatively, the use of a short hydrolysis with hydrochloric acid (see below) may be a useful and less costly way of strengthening the belief that this blue-green-staining matter is DNA.

[*Note:* but remember that DNA stains red after acid hydrolysis.]

On the other hand, most studies on the histochemistry of nucleic acids are likely to be concerned more with whether there is more or less RNA in the cells. For example, it may be required to see whether cells are apparently synthesizing extra amounts of protein, whether they be potentially malignant cells or such cells as lymphocytes responding to produce antibody. Under these circumstances it may be gravely misleading to assume that 'pyroninophilia' equals RNA. The gravity of this assumption can best be emphasized by an example with which we were involved: Histologically the section contained unusual cells which were mitotically active. When tested by the methyl green-pyronin method, these cells stained vividly red and so the suggestion given us was that this was a neoplastic growth. But, when they were tested either by ribonuclease or by the hydrochloric acid method, the ability of these cells to stain with pyronin was undiminished. Consequently the 'pyroninophilia' was not due to RNA. In fact these cells were actively growing, but not malignant, bile duct cells in which the basophilia (to pyronin) was due to some acidic mucopolysaccharide matter. Hence for precise histochemistry, the interpretation of pyronin-positive matter as RNA depends on the loss of this basophilia after treatment either with ribonuclease or with hydrochloric acid.

THE USE OF RIBONUCLEASE

METHOD

This enzyme acts optimally at about pH 7·7. Hence 1 mg of the crystalline preparation is dissolved in 1 ml of 0·2M veronal–HCl buffer at pH 7·7.

[*Note:* it is often far simpler to dissolve 1 mg of the enzyme in an M/40 aqueous solution of sodium bicarbonate: this is a convenient way of achieving a pH of 8·2 which is sufficiently close to the optimum.]

Then: 1. Treat one section with the ribonuclease solution for 1 hr at room temperature.

2. Treat a control section for the same length of time either with the buffer (or bicarbonate) solution alone or with the ribonuclease solution to which has been added zinc sulphate (at a concentration of 0·1–0·01M) to inactivate the enzyme.

3. Wash both sections and stain with methyl green-pyronin.

Result. Pyronin-positive matter, which is lost after treatment with ribonuclease but not after exposure to the control, can be taken to be RNA.

HYDROCHLORIC ACID METHOD FOR RNA

Rationale. Frequently a histochemist may not have ribonuclease readily available but may wish to place more credence on the methyl green-pyronin staining seen in his section than is reasonable by use of the mixture alone. This can be achieved by recourse to the fact that 1 N hydrochloric acid at 60°C extracts all the RNA from sections in 5 min; it also causes sufficient alteration in DNA to make it stain with the pyronin, and not with the methyl green, component of the methyl green-pyronin mixture.

METHOD

Take two cryostat sections. Fix in acetic–alcohol (either 1 : 3 or 5% as required).

1. Take both sections through graded concentrations of alcohol to 0·2M acetate buffer at pH 4·2.
2. Immerse one in 1 N hydrochloric acid which is at 60°C and maintain at 60°C for 5 min.
3. Wash both in 0·2M acetate buffer.
4. Stain both in the methyl green-pyronin mixture (see p. 63).
5. Blot dry; immerse in *n*-butyl alcohol for 2 min. Mount in Euparal (for details see the methyl green-pyronin method as given on p. 63).

Result. RNA stains red in the non-hydrolysed section but is absent from the hydrolysed section. The DNA should stain green-blue in the non-hydrolysed section but a reddish or purple colour in the hydrolysed section.

TOLUIDINE BLUE METHOD FOR BASOPHILIA

Rationale. Toluidine blue is a basic dye and so will stain acidic matter. The exact nature of that matter cannot be determined just by the fact that it stains with this dye; all that can be said is that it is acidic. The use of nucleases can show if it is RNA or DNA. Another diagnostic feature is whether the dye stains orthochromatically, i.e. blue, or metachromatically, namely red or red-purple. (For the histochemical significance of metachromasia see below.)

METHOD

Fresh cryostat sections are recommended. Otherwise fix fresh sections in 1 : 3 (v/v) acetic-alcohol (or 5% acetic acid in absolute ethyl alcohol), and take them through graded concentrations of alcohol to acetate buffer.

1. Stain in an aqueous solution of toluidine blue in acetate buffer for 30 min. [*Note*: 10 min may be sufficient.]
2. Wash in acetate buffer.
3. Blot dry.
4. Place in *n*-butyl alcohol (*n*-butanol) for 2 min.
5. Mount in Euparal.

Solutions required for this method

Acetate buffers (0·1M)

Solution A: 2·85 ml glacial acetic acid, made up to 500 ml with distilled water.

Solution B: 6·8 g sodium acetate made up to 500 ml with distilled water.

For pH 4·2 add 140 ml of A to 60 ml of B
For pH 5·0 add 60 ml of A to 140 ml of B
For pH 5·6 add 20 ml of A to 180 ml of B

Toluidine blue solution

Toluidine blue 0·1 g

Make up to 100 ml with the relevant buffer.

Result. Normal basophilia is shown by blue or purple-blue staining. (The exact hue depends on the composition of the mixture of dyes known collectively as 'toluidine blue'. For this reason it may be preferred to use azure B in place of toluidine blue.) Pink or red staining denotes metachromatic coloration (see below).

METACHROMATIC STAINING WITH TOLUIDINE BLUE

Rationale. There is still considerable uncertainty as to the exact nature of the phenomenon known as metachromasia (e.g. see Baker, 1958, chapter 13; Sylvén, 1954; R. Chayen and Roberts, 1955). The phenomenon itself seems to be quite simple: a pure basic dye in aqueous solutions absorbs light in a distinct manner and so appears to be coloured. [*Note:* the property of metachromasia is seen generally with basic rather than with acidic dyes but is not related, fundamentally, to the acidity of the dye.] When the dye reacts with tissue elements to produce a colour similar to that of the dye, it stains orthochromatically. However certain dyes, even when 'pure', may stain some tissue components orthochromatically but will produce a different colour with other components. As an example we may consider the effects obtained with toluidine blue (which is rarely a simple 'pure' substance but is often a complex of different dyes of different colours, as shown by Ball and Jackson, 1954). This dye, even when it behaves as a single blue dye, stains cells blue (orthochromatically) but collagen red. This change of colour especially when it is shown by a 'pure' dye, is metachromasia; thus toluidine blue stains collagen metachromatically. The change of colour is usually, but not exclusively, towards the longer wavelength (blue to red); it can be shifted towards the shorter wavelengths as, for example, when toluidine blue colours very highly polymerized DNA green.

In complex, but elegant studies, Sylvén (1954) showed that toluidine blue stains metachromatically only when the acidic groups are present in a polymer, and when these groups are closely packed together in the molecule, i.e. they should not be more than about 5 Å apart (0·5 mμ). It may be a fair approximation to suggest that, when the dye reacts with a polymeric molecule in the tissue section (by electrostatic interaction between the basic dye and the acidic group of the polymer), the dye shows its ordinary colour, i.e. it stains orthochromatically. However should these dye molecules come to be so close together along the polymer that they can interact with each other (possibly by Van der Waals' forces), the ionization of the dye becomes suppressed and so gives rise to a change of colour. There seems little doubt from the work of Michaelis (1947) that, spectroscopically, the change of colour is due to the formation of dye micelles (dimers, trimers, probably associated side by side).

To summarize: basic dyes stain acidic substances orthochromatically. They stain metachromatically when the acidic substances are present as long polymeric molecules, with the acidic groups closely packed along the polymer. Consequently acidic mucopolysaccharides are typically stained metachromatically but certain other acidic polymers may also show this type of coloration. The nature of the acidic groups can be tested by finding at what pH value their ionization is suppressed.

METHOD

Fresh cryostat sections should be used (because the state—for example the coagulation—of the cytoplasm can affect the metachromatic response markedly).

Alternatively the sections can be fixed in the picric-formalin fixative.

1. Immerse in an 0·1% solution of toluidine blue at the required pH. Stain for 30 min.
2. Rinse in buffer at the same pH.
3. Either blot dry or shake free of buffer; dry in air.
4. Place in *n*-butyl alcohol for 2 min.
5. Mount in Euparal.

Result. Orthochromatic coloration: blue or blue-purple (depending on the purity of the blue dye).

Metachromatic coloration: red, pink or red-purple.

Solutions required for this test

For pH values of 5·6, 5·0 or 4·2 use the acetate buffer recommended for the usual method for toluidine blue. For lower pH values use distilled water and add a few drops of 1 N hydrochloric acid to achieve the desired pH.

The dye is dissolved (0·1 %) either in buffer or in water. The pH of the dye-in-water solution should be adjusted to the required value after the dye has been dissolved, in case the dye itself should affect the pH.

[*Note:* Azure A can be used in place of toluidine blue. It has been claimed that Azure B (0·1–0·2 mg per ml in citrate buffer at pH 4) stains RNA metachromatically.]

Application. Nucleic acids in sections stain orthochromatically with toluidine blue at pH 4·2. At lower pH values their ability to bind the dye becomes progressively diminished.

Acidic polymers are stained metachromatically by toluidine blue at pH 5·6. If the acidity is due to carboxyl groups, this staining is abolished at lower pH values (e.g. 4·2). Of the acidic groups found commonly in biological material only sulphate groups (and polyphosphate) remain stainable at pH 1·0. Phosphate moieties of organic molecules become less stainable between pH values of 4·2 and 1·0.

EBEL'S TEST FOR POLYPHOSPHATE

Where a region of a section is stained metachromatically at low pH values, the possibility that the acidic groups (responsible for this reaction) may be phosphate can be tested by recourse to this test. Fundamentally it is based on the avidity of lead for phosphate (this is used in the histochemical method for the demonstration of acid phosphatase); the lead which becomes bound to phosphate can be visualized if it is converted into lead sulphide (black or brown).

METHOD (after Ebel, 1952)

Use unfixed cryostat sections.

1. Immerse in 10 % lead nitrate in 0·05M acetate buffer at pH 4·5.
2. Wash well with acetate buffer (or water) to remove excess lead nitrate.
3. Immerse in water which has been saturated with hydrogen sulphide (bubble the gas from a Kipp's apparatus through distilled water for 5 min, preferably in a fume cupboard).
4. Wash in distilled water.
5. Mount in Farrants' medium.

ACRIDINE ORANGE METHOD FOR DNA AND RNA

Rationale. The reaction of acridine orange with both nucleic acids has been much studied. It appears to be a complex situation (reference may be made to papers by Armstrong, 1956; de Bruyn *et al.*, 1953; Massart *et al.*, 1947; Schümmelfeder, 1958; Austin and Amoroso, 1959). Certainly this dye produces a different coloured fluorescence when it is bound to low polymer nucleic acids than when it is linked to high polymer DNA. But it also shows its 'high-polymer' colour when linked to other polymers, such as keratin. The exact colour which is seen depends on the excitation wavelength and on the eyepiece (barrier) filters used in the fluorescence equipment. The action of the dye is also very sensitive to the pH at which the staining reaction is done and to the fixative used.

METHOD (after Armstrong, 1956; Diengdoh, 1966)

Fresh cryostat sections should be used or they may be fixed in acetic–alcohol (1 : 3 v/v). Tissues can be fixed in calcium–formaldehyde but fixatives containing heavy metals should not be used.

1. Immerse in acetate buffer, pH 4·2 for 5 min. [*Note*: this will also fix fresh cryostat sections.]
2. Stain for 15–30 min in acridine orange solution at pH 4·2.
3. Wash in buffer at pH 4·2.
4. Mount in a drop of the buffer and ring the coverslip with wax.
5. Examine with a fluorescence microscope.

Results. (*a*) When excited with purple light, the DNA–acridine orange complex fluoresces green-yellow to bright yellow; the RNA–acridine orange complex gives a dull reddish-brown to flame orange, depending on its concentration in the cells.

(*b*) With ultraviolet light (365 mμ) DNA appears greenish-yellow and RNA crimson red.

'False positives': mast cell granules and cartilage matrix give red fluorescence. Vascular elastic elements fluoresce yellow and keratin is often green.

Solutions required for this reaction

Walpole's acetate buffer at pH 4·2

1 N sodium acetate (13·6 g/100 ml; $CH_3COONa.3H_2O$)	50 ml
1 N hydrochloric acid	35 ml

Acridine orange solution

1 : 10 000 (w/v) parts of acridine orange in Walpole's acetate buffer (pH 4·2).

STAGE 4: REACTIONS FOR POLYSACCHARIDES

Undoubtedly the diagnostic histochemical test for carbohydrate material is the periodic acid–Schiff (PAS) reaction. Even water-soluble carbohydrate matter may be demonstrated provided that it is fixed in the tissue and an alcoholic PAS reaction is used. In plant histochemistry the recognition of the various carbohydrate-containing components depends largely on histological stains (see Table 2). In animal histology, the Alcian blue method is very widely relied upon to demonstrate acidic mucopolysaccharide; Hale's colloidal iron method has also been much used.

TABLE 2

Summary of the properties and colour reactions of cell-wall constituents (modified from Bonner, 1950, *pp.* 130, 131)

Constituent	*Sign of birefringence*	*Ultra-violet absorption*	*Diagnostic colour reactions*
Cellulose	Positive	Little	(*a*) Dichroic violet stain with chlorzinc iodine (*b*) Blue colour with KI—I_2 in conc. H_2SO_4 (*c*) Dichroic stain with Congo red (*d*) Almost no stain with ruthenium red
Pectic substances	Statistically 0	Little	Stain red with ruthenium red
Lignin	0	Strong	Red reaction with phloroglucinol–HCl
Mannans, galactans, etc.	Usually birefringent	Little	No specific stains
Polyuronide hemicelluloses	0	Little	No specific stains
Cuticular substances and 'suberin'	Negative	Strong	Red or orange-red colour with scarlet H or Sudan III
Callose			Stained by resorcin blue (Johansen, 1940)

PERIODIC ACID–SCHIFF (PAS) METHOD

Rationale. Periodic acid is a very strong oxidizing agent which will oxidize a vicinal diol (1,2-glycol), whether in the *cis* or *trans* form, to yield two aldehyde groups:

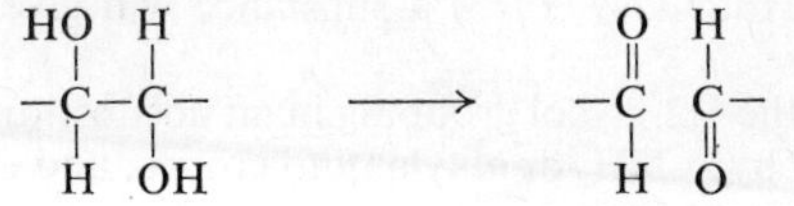

trans-glycol → aldehydes

It should be noted that periodic acid will also oxidize α-amino aldehydes and ketones, α-amino alcohols such as serine, and also diamines although the optimal pH for these oxidations is pH 7–8; pH 3–5 is optimal for the oxidation of α-glycols, hydroxyaldehydes, hydroxyketones and diketones (Dyer, 1956). But the oxidation is inhibited if any of the —OH groups are substituted by any group other than these.

Periodic acid also oxidizes ethylenic linkages (—C=C—) and this is of considerable consequence in histochemistry since these linkages, found in unsaturated fatty acids, may give a strong and apparently specific reaction when so oxidized (see below under benzoylation). Moreover sodium periodate in particular, decomposes readily in sunlight to produce ozone which will oxidize many groups (see Dyer, 1956).

Consequently the object of this test is to use periodate at that concentration, and acting for a suitably short time, which will be most conducive to the oxidation (to the aldehyde) of the —CHOH—CHOH— groups of carbohydrates and not to allow over-oxidation.

The reaction sequence (Hotchkiss, 1948) is as follows:

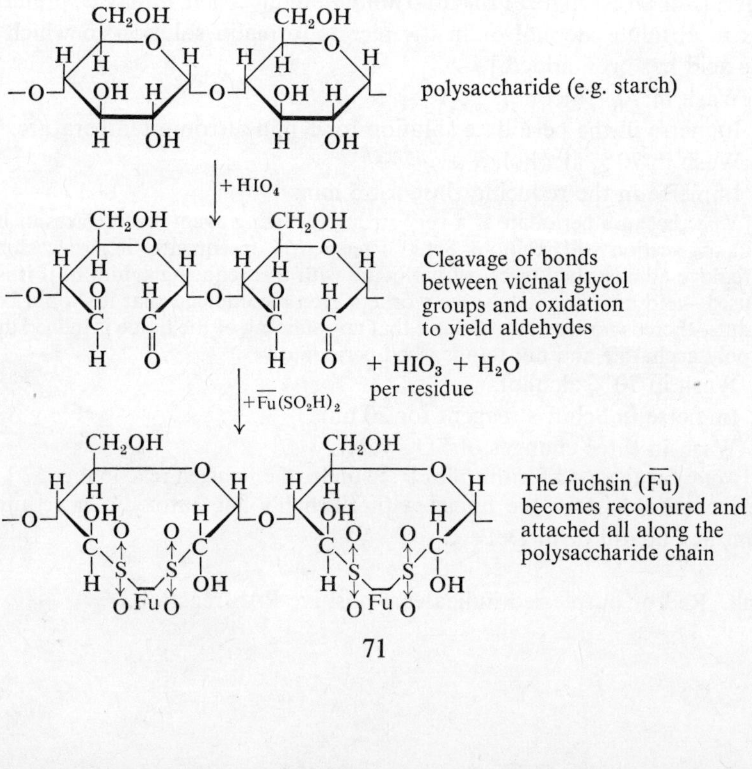

polysaccharide (e.g. starch)

$+ HIO_4$

Cleavage of bonds between vicinal glycol groups and oxidation to yield aldehydes

$+ HIO_3 + H_2O$ per residue

$+ \overline{Fu}(SO_2H)_2$

The fuchsin ($\overline{Fu}$) becomes recoloured and attached all along the polysaccharide chain

According to Hotchkiss (1948) a substance will give a positive result with this reaction if:

(*i*) it contains the 1,2-glycol grouping in an unsubstituted form (except where —OH is replaced by —NH_2 or alkylamino groups, and except for the oxidation products of the glycol grouping);

(*ii*) it does not diffuse away in the course of fixation;

(*iii*) its oxidation product is not diffusible;

(*iv*) it is present in a sufficiently high concentration to give a detectable final colour.

From these criteria it follows that the sugar moiety of the nucleic acids will not be stained (because one of the 1,2-glycol groups is substituted); on the other hand those of cerebrosides and of inositol-lipids will stain by this method.

Considerable controversy has been generated over the problem of how to fix polysaccharides in tissues. The much quoted work of Mancini (1948) which dealt with the diffusion of glycogen in a block of tissue during fixation and dehydration need not affect us here because we can use fresh sections. Polysaccharide material can then be fixed in the section, without diffusion from its site, by the picric-formalin fixative which was developed by Bitensky *et al.* (1962) to deal with the fixation of this type of material. More intense results can often be obtained if 5% acetic acid is added to the fixative, or if the sections are fixed in absolute alcohol; both these procedures produce intensification by partially solubilizing and redistributing the carbohydrate material (see Bitensky *et al.*, 1962).

METHOD (after Hotchkiss, 1948; also see McManus, 1946, for other variants)

Fresh cryostat sections should be fixed for 5 min in the picrate-formalin fixative. [For an estimate of the total amount of glycogen, it may be preferable to fix in absolute alcohol or in the picrate-formalin solution to which 5% acetic acid has been added.]

1. Wash in 70% alcohol.
2. Immerse in the periodate solution for 5 min at room temperature.
3. Wash in 70% alcohol.
4. Immerse in the reducing rinse for 5 min.

[*Note:* because periodate is a very strong oxidizing agent, even traces of it left in the section will recolour Schiff's reagent. Consequently it is necessary to remove all adsorbed traces of periodate with this reducing solution. If it is not used—and unfortunately some workers even recommend that it should be left out—there is no reason to believe that any staining of the tissue is indeed due to polysaccharide and not to adsorbed periodate.]

5. Wash in 70% alcohol.
6. Immerse in Schiff's reagent for 30 min.
7. Wash in three changes of SO_2-water.

[*Note:* for the need for this please see under the Feulgen reaction, p. 57.]

8. Dehydrate (stain the nuclei with Ehrlich's haematoxylin if required); mount in Euparal or in DePeX.

Result. Red or purple-red indicates a positive PAS reaction.

Solutions required for this reaction

Picrate-formalin fixative

40% formaldehyde (technical)	10 ml
Ethyl alcohol	56 ml
Sodium chloride	0·18 g
Picric acid	0·15 g
Distilled water to	100 ml

Periodate solution

Dissolve 400 mg of periodic acid in 15 ml of distilled water. To this solution add 135 mg of crystalline sodium acetate ($CH_3.COONa.3H_2O$) dissolved in 35 ml of absolute ethyl alcohol. Prepare just before use.

Reducing rinse

Potassium iodide	1 g
Sodium thiosulphate	1 g
Distilled water	20 ml

When dissolved, add (while stirring the solution) 30 ml of ethyl alcohol and then 0·5 ml 2 N hydrochloric acid. A precipitate of sulphur may form; this is left to settle out.

Controls. A positive reaction after this procedure could be due to:

1. Aldehydes or ketones present in the tissues and which have remained unaltered during the process;
2. Oxidizable moieties other than carbohydrate; the most obviously significant of these are the $-\overset{|}{C}=\overset{|}{C}-$ groups in unsaturated fatty acids which are oxidized to aldehydes;
3. Carbohydrate matter, probably in a polysaccharide.

To prove that the positive reaction is due to carbohydrate, (or more precisely to —CHOH—CHOH— groups) a control section must be acetylated or benzoylated (see below); matter which is positive to the simple PAS reaction but negative to PAS after such treatment contains the —CHOH—CHOH— grouping and almost certainly contains carbohydrate. The exact nature of the carbohydrate can then be examined by the other staining methods for polysaccharides (starch, glycogen, mucopolysaccharide) or it may be subjected to specific glycolytic enzymes.

To test if some or all of the reaction is due to pre-existing aldehydes or ketones (1, above) another section is subjected to the Schiff's reagent either without prior treatment, or after 5 min in an M/20 alcoholic solution of hydrochloric acid at room temperature (in case plasmalogens are contributing to the response to Schiff's reagent). Should there be pre-existing aldehydes which confuse the true periodic acid-Schiff reaction, they can be blocked by treating the section with a saturated solution of dimedone in 5%

acetic acid for 18 hr at 60°C; subsequent treatment with periodic acid does not uncouple this link between the dimedone and the aldehyde (Gahan, 1965).

BENZOYLATION AND ACETYLATION

(*a*) *BENZOYLATION*

METHOD

1. Fix for 8 min in absolute alcohol.
2. Change to fresh absolute alcohol for a further 8 min.
3. Place in dry acetonitrile in a desiccator containing phosphorus pentoxide for 3 min.
4. Transfer to fresh dry acetonitrile for a further 3 min.
5. Immerse in the benzoylation solution for 2 hr.
6. Remove sections from desiccator.
7. Wash well in absolute alcohol.
8. Continue with PAS test.

Solution required for this method

Benzoylation solution

Acetonitrile	50 ml
Benzoyl chloride	4·2 ml
Dry pyridine	2·2 ml

[*Note:* acetonitrile, benzoyl chloride and pyridine are harmful chemicals and should be handled in a fume cupboard. They should not be pipetted by mouth.]

(*b*) *ACETYLATION*

This is a useful technique in the study of polysaccharides. Occasionally a methylation–saponification method (e.g. Fisher and Lillie, 1954) may be helpful to remove sulphate groups; this procedure and acetylation have been considered by Materazzi and Ferretti (1970).

METHOD

1. Incubate in the acetylation solution at 60°C for 18 hr.
(Occasionally 1 hr may suffice—see Materazzi and Ferratti, 1970.)
2. Wash well in absolute alcohol.
3. Continue with the appropriate test.

Solution required for this method

Acetylation solution

Acetic anhydride	16 ml
Dry pyridine	24 ml

[*Note:* preferably prepare this solution in a fume cupboard.]

'Methylation' (Fisher and Lillie, 1954; Sovari and Stoward, 1970a and b) may sometimes be helpful in analysing the nature of metachromatic material.

ALCIAN BLUE METHOD

Rationale. All the acid glycosaminoglycans (mucopolysaccharides) are negatively charged by virtue of the presence of sulphate ester groups or of carboxyl groups of uronic acids, or of both charged groups. Consequently they bind positively charged (cationic) dyes such as toluidine blue, Azure A and Alcian blue. Of these the last is used very extensively by histologists and histochemists; the many variants can be found in most textbooks of histochemistry. However, the precise use of Alcian blue is that developed by Scott and Dorling (1965) and reference should be made to this work. They applied the concept of the critical electrolyte concentration to differentiate between different glycosaminoglycans. Both in model experiments with chemically defined substances and in histological sections of selected tissues they were able to show that most 'mucopolysaccharides' stained with 0·05% Alcian blue at pH 5·8 (protein interference was noted at pH 2·5); that goblet-cell mucin (mainly carboxyl groups) staining was reduced if the concentration of magnesium chloride was greater than 0·4M; cartilage staining was lost at above 0·6M magnesium chloride; mast cells no longer stained if the $MgCl_2$ concentration was above 0·75M and that of corneal stroma (mainly chondroitin sulphate and keratansulphate) at above 1M. These results agreed with the known concentrations in these tissues of sialo-mucin, chondroitin sulphate, heparin and keratansulphate respectively. But they emphasized that staining with Alcian blue must be carried to equilibrium so that staining should continue for 8 hr. Ideally it should also be shown that the binding of Alcian blue is reversible; this is done by soaking the stained section in the electrolyte solution alone, although de-staining may require longer than staining. However, many workers have reported useful results using only 30 min staining even though this is not rigorously correct (e.g. Andersen *et al.*, 1970).

Some workers have met with difficulties with this method. These difficulties have been clarified by Scott and Mowry (1970). Much of the trouble is that the contents of a bottle, sold under the label of 'Alcian blue', may not correlate well with the label. They listed the following requirements: (1) Use Alcian blue 8 GX (not GS, or 5, or 7 GX); (2) preferably obtain it from I.C.I. Ltd. (Blackley, Manchester); (3) it should dissolve in water to give at least a 5% solution; (4) a 1% (w/v) solution containing 2·0M $MgCl_2$ and buffered at pH 5·7 with 0·025M acetate buffer should not precipitate when fresh, although a fine precipitate may be seen after 24 hr; (5) the spectrum and absorbancies should be as given by Scott (1970); (6) the sample should not be more than 3 years old. With a dye which has these properties, good results should be obtainable.

For details of the nomenclature of mucopolysaccharides see Jeanloz (1960). For histochemical nomenclature and details of other methods see Stoward (1967a, b, c and d).

METHOD (from Scott and Dorling, 1965)

1. Fixation: Scott and Dorling used tissues fixed either in neutral phosphate-buffered 4% formaldehyde or in 95% ethanol at 4°C. Cryostat sections should be fixed; 5 min in the picrate–formalin fixative may be adequate.

2. If necessary take the section through graded concentrations of alcohol to water.

3. Stain at room temperature in 50 ml of the dye solution in a Coplin jar, for at least 8 hr or overnight.

4. Rinse each section individually in a stream of distilled water.

5. Transfer to fresh distilled water in a Coplin jar.

6. Dehydrate through graded concentrations of alcohol (70%, 90%, absolute).

7. Clear in xylene and mount in DePeX.

Result. The various acidic glycosaminoglycans stain blue but this staining can be suppressed by the correct concentration of magnesium chloride.

Solutions required for this method
Dissolve 0·05% Alcian blue in 0·025M sodium acetate buffer at pH 5·8. To stain all acidic glycosaminoglycans add 0·025M magnesium chloride (A.R. grade). This concentration of magnesium chloride can be varied if it is required to demonstrate only particular substances selectively, e.g. use 0·5M $MgCl_2$ to demonstrate chondroitin sulphate and keratansulphate and to suppress sialo-mucins; or 0·7M to stain only keratansulphate.

This solution should be prepared on the day it is to be used.

DIAMINE METHODS FOR MUCOSUBSTANCES

Spicer (1965) developed a range of methods which seem to give valuable information concerning mucosubstances. In particular, a mixture of N,N-dimethyl-*m*-phenylenediamine with N,N-dimethyl-*p*-phenylenediamine (as the hydrochlorides), generally in a proportion of 6 : 1, and at a final diamine concentration of 0·07% at pH 3·4–4·0, may colour acid mucosubstances (sulphomucins and sialomucins) purple. The specificity may be enhanced if iron, as ferric chloride, is included in the dye solution. The 'low iron' diamine method, followed by Alcian blue is said to stain sulphomucins and most sialomucins black; other sialomucins stain blue. A higher concentration of iron and of the diamines, again followed by staining with Alcian blue, is claimed to stain sulphomucins black and sialomucins blue. Gad and Sylvén (1969) tested the high concentrations of iron and diamines on model systems; their results indicated that this method stained sulphomucins provided that red, purple or violet reactions only were taken as truly positive.

To prepare the 'high concentration' dye solution, dissolve 120 mg of the *meta*-diamine and 20 mg of the *para*-diamine in 50 ml of distilled water. Pour into the staining jar which should already contain 1·4 ml of 10% $FeCl_3$. Staining time, for paraffin sections, is 18 hr. The sections can be 'counter-stained' with Alcian blue (1% in 3% acetic acid, for 30 min). The sections can be dehydrated and mounted in the normal way. Gad and Sylvén (1969) used 1·4 ml of 37·2% ferric chloride and adjusted the final pH to 1·3–1·5.

HALE'S COLLOIDAL IRON METHOD (Hale, 1946)

Rationale. This method seems to depend on the adsorption of colloidal iron by tissue components (probably acidic groups); the iron is then stained by the Prussian blue method (see p. 89). There appears to be little reason for believing this method is intrinsically specific for any chemical group but it (or modifications like those of Lillie and Mowry, 1949, or of Wigglesworth, 1952) has been used quite extensively as a histological stain for mucopolysaccharide (or for acidic polysaccharides or for some types of 'mucins').

METHOD

1. Fix cryostat sections in the picrate–formalin solution.
2. Take through graded concentrations of alcohol to water.
3. Flood the section (held horizontally in a staining dish) with the dialysed iron solution. Leave for 10 min.
4. Wash in water.
5. Immerse in acid–ferrocyanide solution for 10 min.
6. Wash in water. (The section may be counterstained with safranin at this stage.)
7. Dehydrate and mount in Euparal or DePeX.

Result. Blue colour indicates 'mucins' or acidic polysaccharides.

Solutions required for this method

Dialysed iron

Pharmaceutical grade of dialysed iron (5% Fe_2O_3)	1 volume
2 N acetic acid	1 volume

Should this not be suitable, or available, the dialysed iron can be prepared fresh (see Gomori, 1952).

Acidic ferrocyanide solution

1% solution of potassium ferrocyanide	85 ml
1 N hydrochloric acid	15 ml

BEST'S CARMINE METHOD FOR **GLYCOGEN**

Rationale. As Gomori (1952) stated, this is an empirical stain for glycogen. There is some doubt nowadays whether it stains all glycogen; it may be sensitive to one form rather than to another (e.g. to the branched rather than to the long chain form of glycogen.)

METHOD*

1. Fix cryostat sections in the picrate–formalin solution for 5 min.
2. Wash in 70% alcohol.
3. Treat with periodate solution for 5 min.
4. Rinse in 70% alcohol.
5. Immerse the section in the reducing rinse for 5 min.
6. Take through graded concentrations of alcohol and stain with Ehrlich's haematoxylin for 15 min.
7. Wash in tap water (or in 5% sodium bicarbonate solution) to 'blue'.
8. Stain with Best's carmine 10–15 min.
9. Wash in the differentiator (a few seconds should suffice).
10. Wash very briefly in absolute alcohol.
11. Clear in xylol and mount in DePeX.

Result. Glycogen should stain red.

Solutions required for this method

Periodate and *reducing rinse*—as for the PAS reaction (p. 73).

Best's carmine

Stock solution

Carmine	2 g
Potassium carbonate	1 g
Potassium chloride	5 g
Distilled water	60 ml

Boil until the colour deepens (3–5 min).
Cool. Add 20 ml concentrated ammonia (0·880).

Staining solution

Stock solution	12 ml
Concentrated ammonia	18 ml
Methyl alcohol	18 ml

Differentiator

Absolute ethyl alcohol	20 ml
Absolute methyl alcohol	10 ml
Distilled water	25 ml

[*Note:* Since the sections are treated with periodate and with the reducing rinse, the results with Best's carmine are exactly analogous to those with the PAS. Hence, in serial sections, those regions which are positive to the PAS reaction and are also stained red by Best's carmine, must be considered to contain glycogen. (This can be proved more rigorously by extraction by diastase or amylase.) If such comparison is not required, steps 3, 4 and 5 can be left out.]

* This method, applied to cryostat sections in this way, has been developed and used routinely by Miss S. Darracott.

LUGOL'S IODINE

This is sometimes of use in polysaccharide histochemistry. Fresh cryostat sections may be used or they may be fixed in the picrate–formalin solution. They are stained for variable periods of time in Lugol's iodine, rinsed and can be mounted in Farrants' medium or dehydrated and mounted in Euparal, etc. Glycogen should stain brownish and native starch should be coloured deep blue. Mancini (1948) dissolved the iodine in liquid paraffin to avoid rendering glycogen soluble.

Lugol's iodine

Iodine	1 g
Potassium iodide	2 g
Distilled water to	100 ml

THE USE OF DIASTASE AND AMYLASES

Suppose that tissue sections contain regions which are positive to the PAS reaction (and perhaps to Lugol's iodine and Best's carmine). The proof that such regions contain glycogen or starch is that they do not give these reactions after specific digestion with a diastase. The simplest way of treating with a diastase is to use saliva but, 'At present, more appetizing and sanitary tests have replaced the time-honoured saliva test' (Gomori, 1952).

METHOD

Fix sections with the same fixative as is used for the PAS reaction. Wash in 70% alcohol and treat the sections for 3 hr at 37°C with an 0·5% solution of α-amylase in 0·004M acetate buffer, pH 5·5. Control sections must be treated with the buffer alone to ensure that any loss of PAS-positive material is due to the specific action of the amylase and is not mere solution of the stainable matter. The sections are then washed in 70% alcohol and subjected to the PAS reaction.

All glycogen and starch will be removed by such treatment. In special investigations it may be found helpful to distinguish between straight-chain and more complex, branched glycogens. To do this, the sections are treated similarly but an 0·5% solution of β-amylase in 0·004M acetate buffer at pH 5·0 is used instead of the α-amylase. The β-amylase digests only straight-chain glycogen (Illingworth *et al.*, 1952).

STAGE 5: REACTIONS FOR LIPIDS (Fig. 11)

Histochemically lipids can be defined as fatty matter which has some of the chemical characteristics of lipid molecules, e.g. they contain choline, or unsaturated fatty acids, or they are phosphatides (for a clear discussion of the chemistry of lipids see Lovern, 1957). Consequently the tests are of two types:

(*a*) Physical methods which depend on the readiness of lipophilic molecules to partition between their solvent and the lipid in the section. These are therefore tests of the degree of 'fattiness' of the material.

(*b*) Chemical methods, like the acid haematein test or the use of Nile blue.

Of these, the physical methods are generally the more important so that, to a large extent, 'lipid' is characterized histochemically by its ability to concentrate Sudan Black or benzpyrene from semi-aqueous solutions in which these compounds are not very soluble.

The histochemistry of lipids has been a very complex and confusing field of study. This is not surprising when it is realized that the biochemical knowledge of these compounds was at best rudimentary before about 1950. There is still a tendency for histologists and histochemists to consider that the term 'lipid' is synonymous with 'fat'—by which is meant fat droplets which may range from large, microscopically obvious droplets or globules (most strikingly seen by phase contrast microscopy of fresh cryostat sections) down to minute droplets dispersed throughout the protoplasm and not resolvable by the optical microscope. The latter form of fat droplets were sometimes referred to as 'masked lipids' but this term is to be avoided for this type of fat since it is not 'masked'; it is merely finely dispersed (see Berg, 1951; Chayen *et al.*, 1959). The term 'masked lipids' should be restricted to those lipids which are actually, and usually chemically, masked by being covered by (or bound to) some other material, such as protein or carbohydrate. It is this form of lipid which plays an essential role in the structure of cells (and not only in the membranes), and which does not react with lipid colourants or reagents until the masking molecule has been displaced or disjoined from the lipid. It is still not generally appreciated that a lipid-protein complex (like that isolated by Green, 1959, from the hydrogen-transport matter of mitochondria) may contain 96% of lipid and yet be water-soluble, i.e. it is not 'fatty'. Confusion has arisen, therefore, because preparatory or staining procedures, or physiological variations, may themselves cause more or less unmasking of structural lipids (held as complexes in the protoplasm). Consequently, workers have often reported methods as being capricious when in fact it is not the method which is variable, but the degree to which the lipid is bound to its masking substance. In fact, the changes in this binding induced by physiological variations, by trauma, drugs and hormones provide important and sensitive indications of the function of the tissue (see Chayen, 1968, for a review of this subject).

SUDAN BLACK B METHOD

Rationale. Sudan black B is a fat-soluble colourant which is soluble in absolute alcohol but insoluble in water. It is a strong colourant, and has a high affinity for fatty material, including phospholipids (see Baker, 1958, pp. 300–1). It is used as a saturated solution in 70% ethyl alcohol in which it is still soluble (it is very much less soluble in 50% alcohol) and the section is exposed to this solution. The colourant partitions between the alcohol and the tissue lipids (in which it is much more soluble). Because of its intense colour, and of its strong affinity for all (or most) classes of lipids, the colouration of tissue components by Sudan black B is often taken as the histochemical criterion for the presence of lipids. The purpose of the various 'burnt Sudan black' methods is to increase the sensitivity of tissue lipids, including some masked lipids but also steroids (which, being solid, will stain only at higher temperatures, i.e. closer to their melting point). The drawback of the procedure is the fact that the section is exposed to 70% alcohol which may itself affect the lipids.

METHOD

Fresh cryostat sections must be used because most, if not all, fixatives can modify the nature of lipid-protein and lipid-carbohydrate complexes.

1. Place the section horizontally in a staining tray and filter the Sudan black solution through a funnel on to the section. Leave for 30 min (replenishing if necessary).
2. Remove any deposit of the Sudan black from the section by rinsing briefly in 70% ethyl alcohol. Transfer to 50% alcohol to judge if the deposit has been adequately removed.
3. Wash in 50% alcohol.
4. Wash in water.
5. Mount in Farrants' medium.

Result. All available lipids (other than steroids) should be coloured grey to blue-black.

Solutions required for this method

A saturated solution of Sudan black B in 70% ethyl alcohol. This can be kept for prolonged periods, provided that the solution is filtered on to the section.

[*Note:* Some workers prefer to use a 0·7% solution of Sudan black B in propylene glycol and to differentiate in 85% propylene glycol. Gomori's (1952) suggestion that the dye should be dissolved in 60% triethylphosphate is interesting but it seems not to be much used.]

MODIFIED SUDAN BLACK METHODS

The 'burnt Sudan black' method recommended by Burdon (1947) for use with micro-organisms sometimes can be remarkably helpful when applied to tissue sections, especially if the presence of steroids is suspected. The procedure is the same as for the ordinary technique except that, when the solution of Sudan black B has covered the section, the alcoholic solution is set alight. When the alcohol has finished burning, the section is covered with fresh solution and the process is repeated a few times. Care must be taken to remove the heavy deposit of solid Sudan black with 70% alcohol.

Berenbaum (1958) has suggested other ways by which tissue lipids may be unmasked and coloured with Sudan black B.

OIL RED O METHOD FOR COLOURING FATS

Rationale. Oil red O is a strong colourant of fats but it requires a high concentration of fatty matter for it to be drawn out of its alcoholic solution into the section. Triglycerides ('fats') are 'fattier' than are phospholipids which contain polar groups (i.e. hydrophilic and so lipophobic parts of the molecule); moreover the former occur more generally in highly concentrated globules of hydrophobic molecules whereas the latter may be associated with other more hydrophilic molecules. Consequently oil red O is an excellent colourant of fats but very poor for phospholipids.

METHOD

Fresh cryostat sections must be used.

1. Rinse the section in 60% isopropyl alcohol.
2. Immerse in oil red O solution for 10 min.
3. Rinse briefly in 60% isopropyl alcohol.
4. Wash in water; then either mount in Farrants' medium or proceed to 5.
5. If required, stain in Mayer's haemalum or in Harris' haematoxylin for ½–1 min; wash in water; blue in 5% sodium bicarbonate for a few minutes; wash well in distilled water and mount in Farrants' medium.

Result. Fats are stained red or yellow-red, depending on their concentration.

Solutions required for this method

Stock solution of oil red O

A saturated solution (*c.* 0·5%) of oil red O or of oil red 4B in absolute isopropyl alcohol. This solution can be kept indefinitely.

'Staining' solution

Stock solution of oil red O	60 ml
Distilled water	40 ml

Leave to stand for 10 min and then filter before use.

[*Note:* this solution is over-saturated (so as to make the oil red O more likely to dissolve out of this solution and into the fat). Consequently it must be filtered just before use (cf. the Sudan black method).]

NILE BLUE METHOD FOR LIPIDS

Rationale. This technique has been investigated fully by the Oxford school of cytologists (reference should be made to Baker, 1958, pp. 301–3). An aqueous solution of Nile blue A contains the blue cation of the dye; the red spontaneous oxidation product of it, namely Nile red; and the orange-red imino-base of Nile blue. The Nile red dissolves in all fatty lipid components and acts as a general lipid colourant. Both the cation of the Nile blue, and the imino-base can react with acidic groups and when so linked they colour blue. Consequently phospholipids and fatty acids will stain blue (because these components of the Nile blue solution will overlay and mask any effect of Nile red) but so may other non-lipid acidic groups; the blue colour is specific for fatty acids and phospholipids only if the region so coloured has been proved to contain lipid. On the other hand, neutral fats will be coloured red. The difficulty in the past has been that the amount of Nile red present in any given solution may vary considerably. This is overcome either by adding Nile red (which can be obtained commercially) to a solution of Nile blue A, or by using commercial mixtures which have been reinforced with the red dye.

METHOD

Cryostat sections must be used.

1. Warm all solutions (i.e. 1% aqueous Nile blue, 1% acetic acid and water for washing) to either 37°C or 60°C before beginning the reaction.
2. Immerse the sections in a 1% aqueous solution of Nile blue at 37°C or 60°C for 5 min.
3. Wash quickly in the warmed water.
4. Immerse for 30 sec in warmed 1% acetic acid.
5. Wash in warmed water and mount in Farrants' medium.

Compare this section with a second section which is stained and differentiated the same as the first and then re-stained with an 0·02% aqueous solution of Nile blue at 37°C or 60°C for 5 min. Then wash quickly in warmed water; differentiate for 30 sec in warmed 1% acetic acid; wash in warmed water and mount in Farrants' medium.

Solutions required for this method

1% *Nile blue*

Nile blue	1 g
Distilled water to make up to	100 ml

0·02% *Nile blue*

Nile blue	20 mg
Distilled water to make up to	100 ml

BENZPYRENE METHOD FOR LIPIDS

Rationale. As has been discussed (in relation to the use of Sudan black), the colouration of lipid matter by lipophilic dyes depends on:

(*i*) how soluble the dye is in its solvent as against its solubility in the lipid (i.e. the partition coefficient of the dye in the lipid and the solvent). It would be ideal if the dye could be 'dissolved' (in sufficient concentration) in a solvent in which it was virtually insoluble because then it would be more readily attracted to the tissue lipid. Moreover, a good solvent for the dye (such as absolute alcohol) is likely to be a solvent for the tissue lipid as well.

(*ii*) the intensity of the dye itself: a strongly coloured dye will be detected even in relatively low concentrations in the section.

The benzpyrene method (Berg, 1951) seems to meet both requirements better than any other technique. Benzpyrene is an intensely fluorescent molecule so that it meets requirement (*ii*). Furthermore it is possible to detect far less of a fluorescent dye than of a visible dye (see Chayen and Denby, 1968) so that the sensitivity of this method should far exceed that of the others discussed in this section. The other special advantage of benzpyrene is that it can be dispersed in a 'hydrotropic solution' in water if this dispersion is stabilized by caffeine. It therefore approximates well to the first requirement (above).

In our experience it is indeed a very sensitive method for all types of lipids (except solid steroids). It can also be used as a stain for living cells (e.g. Dumonde *et al.*, 1965).

METHOD

Fresh cryostat sections should be used although the technique works on fixed sections, even if treated with dichromate (as in the acid haematein procedure).

1. Cover the section with the benzpyrene reagent for 20 min.
2. Rinse well.
3. Mount in water and examine with the near ultra-violet light (e.g. 365 mμ) with a fluorescence microscope.

For cell suspensions. Add 0·6 ml of the benzpyrene reagent to about 1 ml of the cell suspension. Leave for 20 min. Make a smear or hanging-drop preparation (preferably in fresh suspending medium) and inspect with the fluorescence microscope.

Result. Lipid fluoresces silver to blue-white or even yellow, depending on the concentration of the benzpyrene in the lipid.

Control. Use the caffeine solution without added benzpyrene.

Solutions required for this method

To prepare the benzpyrene reagent (Berg, 1951):

1. Saturate distilled water with caffeine at room temperature for 2 days (about 1·5% caffeine).
2. Filter.
3. Add about 2 mg of 3,4-benzpyrene (also called benzo-pyrene) to 100 ml of this caffeine solution.
4. Incubate at about 35°C for 2 days.
5. Filter off the excess benzpyrene.
6. Dilute the solution with an equal volume of distilled water (to prevent the caffeine from crystallizing out of the solution).
7. Leave to stand for 2 hr.
8. Filter.

This solution is now ready for use although it can be kept for several months in a dark bottle. It should contain about 0·75 mg of benzpyrene in 100 ml of solution.

Rapid method for preparing the benzpyrene reagent

1. Prepare a 1% solution of benzpyrene in acetone.
2. Add this, drop by drop, to 100 ml of a saturated solution of caffeine which must be stirred briskly. When a persistent turbidity is obtained, enough benzpyrene has been added. (0·1–0·2 ml has been recommended, but we have found rather more to be required.)
3. Dilute with an equal volume of distilled water.
4. Filter twice.

The solution is then ready for use.

Notes: (a) *Great care should be taken in handling the concentrated solutions of benzpyrene because this substance is* **carcinogenic.**

(b) In practice we prefer the slower method of preparing this reagent.

OTHER FLUORESCENT METHODS FOR LIPIDS

Rhodamine B (1:1000 or even 1:10,000 in water, or in some Ringer solution) has been much used as a vital fluorescent dye (fluorochrome), particularly for plant cells (see Strugger, 1938; Höfler, 1949). It is quite a useful colourant of lipids. The phosphine 3R method of Volk and Popper (1944) is another useful fluorescence method for lipids.

THE ACID HAEMATEIN METHOD FOR **PHOSPHOLIPIDS**

Rationale. This whole subject has been reviewed fairly extensively by Baker (1946) and by Chayen (1968). In brief it is claimed that this method is specific for phospholipids and can yield valuable information regarding the state of these molecules inside the tissue (as suggested above).

In the procedure, the sections are treated with dichromate first at room temperature and then at 60°C. Particularly at the higher temperature the chromate reacts with the unsaturated bonds of the fatty acids of phospholipids (but not generally of fatty acids alone or in triglycerides, as shown in Chayen, 1968). The chromium becomes very tightly bound to the phospholipid. The section is then exposed to acidified haematein which forms a metal–dye complex (a 'lake' in colourists' terms). This 'lake' occurs wherever chromium occurs in the section. The tissue is then immersed in borax–ferricyanide which dissolves the haematein out of the section; it is less readily removed from the phospholipids because the chromium is linked to the phospholipid as well as to the haematein. The proof that this retention of the haematein is caused by unsaturated bonds is proved by the use of control sections in which all these bonds are brominated (and so saturated) before the acid haematein reaction is done (the sections should be negative under these conditions).

Calcium is added to the dichromate solution to render the lipids less soluble than they might be in the dichromate solution alone (see Baker, 1946).

Some workers have been worried that the acid haematein method can give variable results. In many, if not all cases, these variations have been due to various degrees of unmasking of the phospholipids either in life, or as a result of the way the tissue was treated. For example, it is not widely appreciated that the hot dichromate itself can cause some degree of unmasking. A more general cause for concern was the fact that lipid solvents often enhanced, rather than eliminated, this reaction. It is now known that these solvents, acting for relatively short periods, are the most effective unmasking agents available; with prolonged treatment they will remove the fully unmasked lipid.

METHOD

Fresh cryostat sections must be used for this method. They must be taken straight from the cryostat and used immediately.

1. Immerse in dichromate–calcium solution in a Coplin jar at room temperature for 18 hr.
2. Transfer the Coplin jar to an incubator at 60°C and leave for 24 hr.
3. Wash well in distilled water; use several changes to remove free dichromate.
4. Immerse the section in acidified freshly prepared haematein for 5 hr at 37°C.
5. Rinse in distilled water.
6. Place the section in the borax-ferricyanide solution and leave for 18 hr at 37°C (in the dark).
7. Rinse in distilled water.
8. Then: either take the section up through graded concentrations of alcohol and clear in xylol, or shake off the water; dry the section in air; rinse in *n*-butanol.
9. Mount in Euparal.

Result. Phospholipids stain various shades of blue and black. Some types of lipid may stain deep brown. This is not a spurious reaction, as has been shown by studies with purified phospholipid material. The section may appear yellowish but this is a general non-specific colouration.

Solutions required for this method

Dichromate–calcium solution

Potassium dichromate	5 g
Calcium chloride (anhydrous)	1 g
Distilled water to make up to	100 ml

Acidified haematein solution

1. To 0·05 g haematoxylin (B.D.H. 'SS Reagent' grade) add 48 ml of distilled water and 1 ml of a 1% solution of sodium iodate.

2. Heat until the solution boils. [*Note*: this oxidizes and so 'ripens' the haematein, as can be seen by the deepening colour.]

3. Cool and then add 1 ml of glacial acetic acid. This solution should be prepared immediately before it is to be used.

Borax–ferricyanide solution

Potassium ferricyanide	0·25 g
Sodium tetraborate. 10 H_2O	0·25 g
Distilled water to make up to	100 ml

This solution should be kept in a dark bottle.

To unmask the phospholipids (and so to disclose the total concentration of these substances in the section):

Treat sections with 1 : 2 (v/v) methanol–chloroform for 1–10 hr at room temperature. Generally, optimal unmasking of the phospholipids occurs after between 3 and 5 hr; prolonged treatment (up to 24 hr) extracts the lipids. Stain with the acid haematein method.

Bromination as a control for the acid haematein method

1. Unmask phospholipid by treatment for 3–5 hr with methanol–chloroform (1 : 2, v/v).

2. Immerse in a saturated solution of bromine in water for up to 24 hr.

3. Rinse and test with the acid haematein method.

To prepare a saturated solution of bromine, add 1 ml of liquid bromine to about 50 ml of distilled water in a Coplin jar. Stir well. It is advisable to keep the jar closed whenever possible.

METHODS FOR STEROIDS

As was discussed by Cain (1950), because steroids are solid and often crystalline, they cannot be coloured by lipid dyes unless the reaction is done at a temperature at which the steroids melt (see 'burnt Sudan black' method above). They may sometimes be detected in a polarized light microscope by their birefringence (see Adams, 1965). It has been claimed that they can be demonstrated by their fluorescence: as in the method used in chromatography, steroids may show primary fluorescence but they show greatly enhanced fluorescence (sometimes with striking change of colour) when subjected to concentrated sulphuric acid. This can be done with cryostat sections provided that they are dried in a vacuum desiccator and are not removed from the desiccating environment until the section has been covered with the sulphuric acid (Chayen *et al.*, 1966). Another useful method for steroids is the microchemical Liebermann-Burchardt reaction. In this (Gomori, 1952; Schultz's method; also see Cain, 1950 for fuller discussion) cryostat sections are treated with 2% ferric alum for 24 hr. They are then rinsed and mounted in a freshly prepared mixture which contains equal volumes of glacial acetic acid and concentrated sulphuric acid. The slide must be inspected immediately: colours change from purple to dark blue or to blue-green. The last colour indicates the presence of unsaturated steroid such as cholesterol and its esters. A potentially valuable and more truly histochemical technique for steroids is that described by Adams (1961). New methods for ketosteroids have been described recently (Adams, Smith and Stoward, 1967).

METHODS FOR **CAROTENOIDS**

Probably the most useful way of demonstrating these substances is by their yellow to green autofluorescence. Although this fluorescence is often said to be only transitory, it has been found to be remarkably permanent when vitamin A has been dissolved in lipid, and for the carotenoids of the retina, studied in cryostat sections (Chayen *et al.*, 1966).

STAGE 6: MINOR COMPONENTS (Fig. 12)

INORGANIC ELEMENTS

There are a number of well documented methods for detecting relatively large concentrations of inorganic elements in tissue sections.

IRON (see Casselman, 1959)

(*a*) *Unmasked iron*, as in haemosiderin: the various modifications of Perls' methods are generally recommended. Fresh cryostat sections can be used. It is advisable to avoid acidic fixatives which may allow the iron to diffuse; Casselman preferred neutral formalin to alcohol.

METHOD

1. Immerse the section in freshly prepared, filtered 2% potassium ferrocyanide solution for 5 min.
2. Transfer to acidified ferrocyanide solution (equal volumes of the 2% potassium ferrocyanide solution and 0·2 N hydrochloric acid); leave for 20 min.
3. Wash in distilled water; nuclei can then be stained with neutral red.
4. Dehydrate and mount in Euparal.

Result. Reactive (unmasked) iron stains blue.

(*b*) *Masked iron*, as in haemoglobin; treat the section for at least 30 min with 100 volume hydrogen peroxide to which has been added a few drops of sodium carbonate solution, to render it slightly alkaline. Then stain for reactive iron as above.

CALCIUM

Casselman (1959), who based his views largely on the work of McGee-Russell, recommended either the nuclear fast red or the alizarin sulphonate method (see McGee-Russell, 1955, 1958). Although the latter is by no means specific for calcium, it may be useful when calcium is suspected in the absence of high concentrations of some less common metallic ions which otherwise can confuse the issue.

Puchter *et al.* (1969) have contributed a thoughtful and detailed review of the mode of action of alizarin compounds and recommend the use of alizarin red S at pH 9.

PIGMENTS

Many histochemical pathologists have paid considerable attention to tissue pigments. The chemistry of these substances seems less important, and less known, than is their histological terminology and reference should be made to Gomori (1952), Pearse (1960) or Barka and Anderson (1963).

REDUCING SUBSTANCES

Many reducing substances can be observed in tissues by their reduction of a 1% aqueous solution of osmium tetroxide or of various silver nitrate solutions. Ascorbic acid is so strong a reducing agent that it also reduces acidified silver nitrate (for full review see Chayen, 1953); others can act only on alkaline or ammoniacal solutions of silver nitrate (e.g. the 'argentaffin reaction' in which polyphenolic substances reduce ammoniacal silver nitrate to metallic silver).

Phenolic substances, especially polyphenols, are of some importance in histochemistry since they include adrenaline and enterochromaffin substance (Gomori, 1952). All phenols seem to react with ferric chloride to yield pale colours which may be characteristic of the phenol. Diphenols in particular are oxidized by dichromate to form coloured quinones (the 'chromaffin reaction'). No fixation is required because the dichromate is a mild fixative itself, so that the reaction and fixation occur simultaneously in cryostat sections.

AUTOFLUORESCENT SUBSTANCES

Certain molecules, when they are irradiated with light of short wavelength, have the ability to absorb this light-energy (the excitation light) and to transmit some of the absorbed energy as light of longer wavelength (transmitted light). This is what is meant by fluorescence. This should be of considerable value in histochemistry because a number of physiologically and pharmacologically significant molecules are capable of being rendered fluorescent (i.e. they show 'autofluorescence'); many of these molecules cannot be identified adequately by any other histochemical method. In view of this, it seems surprising that so little use has been made of primary fluorescence microscopy (namely the microscopic study of autofluorescent molecules; this is contrasted with secondary fluorescence, in which the fluorescence is induced by the use of fluorescent dyes or fluorochromes, such as benzpyrene). To a large extent the little reliance that has been placed on fluorescence microscopy has been caused by the fact that many factors, including certain fixatives, fat-solvents and heat, cause tissue to become intensely and apparently non-specifically autofluorescent. Some of this 'non-specific' autofluorescence may be caused by lipid peroxidation induced by the splitting of lipid-protein complexes by these agents. Other workers have been concerned that autofluorescence can be 'quenched' (i.e. the molecule ceases to emit light) by a wide range of agents, such as metals (used in some fixatives) and oxidative or reductive media in which the cells may be mounted.

Most of these objections are eliminated by the use of unfixed cryostat sections which can be mounted either dry in air, or in liquid paraffin (which can have a high concentration of oxygen) or in any medium which is especially suitable for the particular autofluorescent molecule to be studied. Some substances which can be identified by fluorescence microscopy are listed in Table 3. The application of this form of microscopy has been reviewed recently by Chayen and Denby (1968).

TABLE 3

Substance	*Fluorescence*
Riboflavin	Yellow—in 1 N acetic acid it fluoresces intensely yellow-green when irradiated with ultra-violet light
Porphyrins	Red
Thiamine	White
Pteridines (including folic acid)	Generally blue-white, but other colours have been reported
Ceroid and lipofuscins	Yellow or brown

INDUCED FLUORESCENCE OF CATECHOLAMINES AND RELATED SUBSTANCES

Over the past few years much use has been made of formalin-induced fluorescence for detecting catecholamines and related compounds in cells (e.g. Eränkö, 1967). This appears to be a growing interest in certain specialized branches of histochemistry and reference should be made to the original papers by those wishing to utilize these very sensitive techniques. In general monoamines, kept at their original sites by freeze-drying or in cryostat sections (Hamberger and Norberg, 1964; Eränkö and Härkönen, 1965), are treated with slightly humid gaseous formaldehyde (from paraformaldehyde) at 80°C for 1 hr to convert them into a condensation product (tetrahydro-isoquinoline derivative). This substance undergoes dehydrogenation, through the mediation of dry protein, to form a highly fluorescent dihydroisoquinoline derivative (Corrodi and Jonsson, 1967).

R–(phenylethylamine)–NH_2 + HCHO → R–(tetrahydroisoquinoline)–NH → R–(dihydroisoquinoline)=N

dihydroisoquinoline
very fluorescent

condensation (can occur readily in solution) →

dehydrogenation (occurs only in relatively dry films) →

Corrodi and Jonsson showed that as little as 0·001% w/v of primary catecholamines or 5-hydroxytryptamine can be demonstrated in a 10 μ section. In investigations of an adrenergic nerve cell body of the superior cervical ganglion in the cat it has been calculated that the fluorescence method detected $0{\cdot}4 \times 10^{-6}$ μg of noradrenaline and similar results have been obtained in other cells with dopamine. Particularly if quantitative microspectrofluorimetry is used, it is possible to distinguish between catecholamines and 5-hydroxytryptamine; the emission spectrum of the fluorescence of the former showing a maximum at 480 nm and of the latter at 525 nm. Spectral changes can also be induced by exposure of the fluorescent compounds to dry hydrogen chloride gas. Björklund *et al.* (1968) used such treatment and showed that the fluorescence excitation spectra of dopamine and noradrenaline became so different that the two substances could be clearly differentiated. Incorporating minute amounts of the dry hydrogen chloride gas into the formaldehyde vapour appears to increase the amount of fluorescence which is inducible, and so makes it possible to detect compounds whose fluorescence is too weak when exposed to formaldehyde alone (Björklund and Stenevi, 1970). For example, the fluorescence of tryptamine was 20–200 times greater with acid treatment than with the standard formaldehyde method. They suggested that the acid catalysed the condensation step.

The specificity of the induced fluorescence can be tested by treating the sections with low concentrations of sodium borohydride ($NaBH_4$) in alcoholic solution. This instantly reduces the fluorescent compounds to non-fluorescent molecules which become fluorescent again after renewed treatment with the formaldehyde gas.

There is doubt whether histamine can be demonstrated by these methods (Corrodi and Jonsson, 1967). However, it can be demonstrated by its reaction with *o*-phthalaldehyde, either in an anhydrous solvent or in gaseous form, to form a highly fluorescent compound (Shelley *et al.*, 1968). The reagent can also be used as a supravital dye (Juhlin, 1967).

AUTORADIOGRAPHY*

Autoradiography allows the localization of radioactively labelled material in a specimen by means of the photographic action of the emitted ionizing

* We are grateful to Dr. S. R. Pelc for his help with this section.

particles. In all techniques the specimen is brought into contact with a photographic emulsion and stored in the dark for a certain time, the exposure time. The film is then developed and fixed and can be examined macroscopically or through a microscope. With low or no magnification, usually called macroautoradiography, the presence of radioactive material is indicated by a blackening of the photographic emulsion; at higher power the blackening is seen to be produced by a larger or smaller number of black particles, the photographic grains. Under the electron-microscope the grains are seen to be filamentous.

TABLE 4

	Autoradiographic methods for: *Fixed compounds*	*Soluble compounds*	*Resolving power*
Macro	Apposition	Ullberg (7)	15–50 μm
Light-microscopy	Stripping film method (1) Dipping method (2)	Appleton (8) Stumpf and Roth (9)	1–2 μm for H^3 2–5 μm for other isotopes
Electron-microscopy	*see* Salpeter and Bachmann (3) Budd and Pelc (4, 5) Caro (6)	No technique available	0·1–0·3 μm

References to Table 4

(1) Pelc, S. R. 1956. *Intern. J. Appl. Radiation Isotopes* **1**, 172. (2) Joftes, D. L., and Warren, S. 1955. *J. Biol. Phot. Assoc.* **23**, 145. (3) Salpeter, M. M., and Bachmann, L. 1964. *J. Cell Biol.* **22**, 469. (4) Pelc, S. R., Coombes, J. D., and Budd, C. C. 1961. *Exptl. Cell Res.* **24**, 192. (5) Budd, C. C., and Pelc, S. R. 1964. *Stain. Technol.* **39**, 295. (6) Caro, L. G. 1962. *J. Cell Biol.* **15**, 189. (7) Ullberg, S. 1962. *Biochem. Pharmacol.* **9**, 29. (8) Appleton, T. C. 1964. *J. Roy. Microscop. Soc.* **83**, 277. (9) Stumpf, W. E., and Roth, L. J. 1966. *J. Histochem. Cytochem.* **14**, 274. Full details of these references and of these techniques will be found in Rogers (1967) and Gahan (1972).

The introduction of labelled material into an organism is usually in soluble form and the aim of an experiment can be to localize it in its original state (e.g. Na^+ ions) or after utilization as a precursor (e.g. labelled amino-acids in protein-synthesis or H^3-thymidine in DNA-synthesis). Many soluble compounds are lost or moved through diffusion during fixation, washing, dehydrating and embedding. Therefore autoradiography of a fixed and sectioned specimen can only be attempted when insoluble labelled substances are to be scored and the loss of labelled precursor is regarded as an advantage. When soluble material is to be retained frozen sections are used. The techniques serving various combinations are shown in Table 4.

Certain practical points have to be noted.

(*i*) The particles emitted by the isotope have different energies and therefore different ranges. The worst possible resolving power is determined by the range, which is of importance when ^{3}H is used whose β-particles are so weak, i.e. their range is so small, that resolving powers of 1 μm are easily achieved. The accompanying disadvantage is self-absorption within the specimen such that only β-particles emitted within a short distance from the emulsion are able to reach it (Fig. 14).

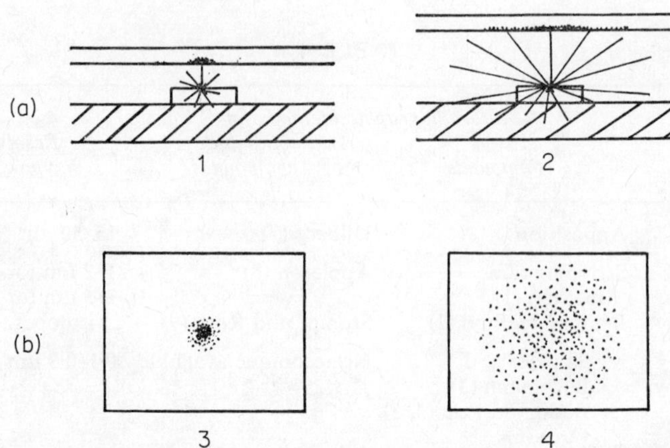

Fig. 14. Diagrammatic representation of stripping film autoradiography (a) and (b) of the image of a point source of radiation in the specimen (x) as seen on the photographic film when there is little gap (1 and 3) or a large gap (2 and 4) between the emulsion and the specimen. The point source of radiation will be fairly well localized if the photographic emulsion is closely applied and poorly localized if the film and section are widely separated (also see Fig. 15). The strength of the radiation will also affect the localization because stronger radiation penetrates further than does weaker radiation.

(*ii*) The geometry of emulsion and source of radiation are important for emitters of high energy β-particles. As a general rule it can be assumed that the resolving power is one-half of the combined thickness of specimen and photographic emulsion if there is no gap between them. For example, a 3 μm section with a 1 μm emulsion gives 2 μ resolving power. The effect on resolution of the gap between the emulsion and the specimen is shown in Fig. 15.

(*iii*) The size of the photographic grains limits the resolving power since obviously no detail smaller than the grains can be resolved. The usual grain sizes are between 0·1 and 0·25 μm and do not materially affect the resolving power for light-microscopy but become important for electron-microscopy work.

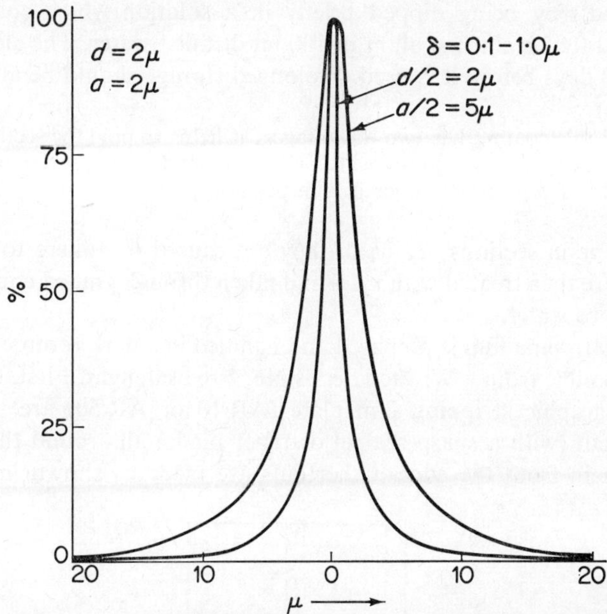

Fig. 15. The calculated relative distribution of grains produced by a radioactive source at 0 in a section of thickness (d) of 2 μ and with an emulsion of thickness (a) of 2 μ. The size of the gap between emulsion and section (δ) is 0·1 μ in the inner curve and 1·0 μ in the outer curve. The resolution is defined as the width of the distribution-curve at the half-way position ($d/2$). (From Doniach and Pelc, 1950. We are grateful to the *British Journal of Radiology* for permission to use this figure.)

Autoradiographic methods and uses

The various methods of autoradiography are now so valuable and so widely used that a wide range of different techniques is available.

For various methods, and some of the uses of autoradiography, reference should be made to Pelc (1958), Taylor (1956), Lima-de-Faria (1961), Fitzgerald (1959), Rogers (1967) and Gahan (1972). The basic stripping-film technique is as follows:

1. *Sections.* For most studies, which have been concerned with substances which can be fixed in cells, such as DNA or protein, fixed tissues have been embedded in paraffin and sectioned normally but smears or squashes can be used. Certain metals, used in some fixatives, themselves react with photographic emulsions; consequently 1 : 3 acetic : alcohol, or formaldehyde are usually the recommended fixatives. Cryostat sections of unfixed tissue can be used, but should be fixed in acetic-alcohol or exposed at −25°C.

[*Note:* for water-soluble material it is essential to use unfixed cryostat sections.]

2. The *slides* on which the sections are to be picked up from the knife must be washed well in hot water (but not treated with acid) before being coated

(or 'subbed') by being dipped briefly in a solution which contains 0·5 g gelatin and 0·05 g chrome alum in 100 ml distilled water. The slide must be dried for 2 days before it is used. Prolonged storage should be at +4°C in a refrigerator.

[*Note:* this coating has two advantages: it helps to hold the section on to the slide during all the subsequent processing and it stops the photographic emulsion from slipping once it is in position.]

3. If paraffin sections are used, they are caused to adhere to the coated slide and are then treated with xylol and taken through graded concentrations of alcohol to water.

4. The stripping film is prepared and handled in a dark room, with as little light as possible from a Wratten Series No. 1 red safelight. First, the edges of the photographic stripping film plate (AR 10 or AR 50) are trimmed by tracing a cut (with a sharp scalpel or razor blade) all around the plate and about ½ in in from the edge. Other cuts are made as shown in Fig. 16 to produce eight segments of film. Note that stripping film is a double layer of pure gelatine (nearest the glass) and emulsion. The emulsion must be nearest the specimen.

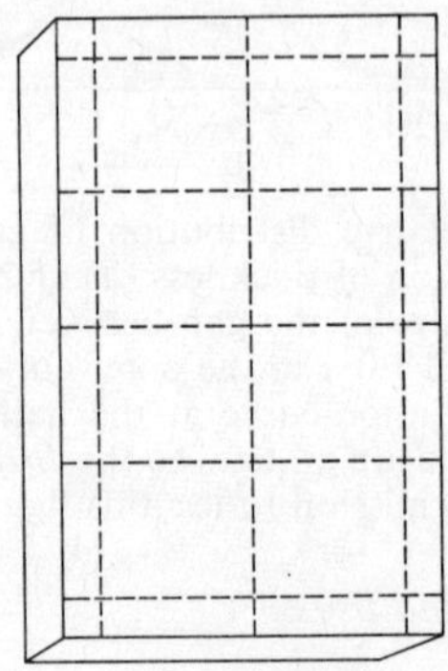

Fig. 16. Diagrammatic view of a stripping film on its plate-backing. Cuts (broken lines) are made in the photographic film to produce eight segments of film.

With the scalpel pick up the edge of each segment of film and very slowly strip that segment off the plate. Hold it with as little contact as possible and float it, face (emulsion) downward, on to the surface of water in a large bowl. The temperature of the water should be maintained at 21–23°C to permit the maximum expansion of the film (2½ min). Six to eight pieces of film can be floated out together.

5. Place the slide, section uppermost, under the water and slowly raise it, under the film. Ideally the slide should be held slightly oblique to the film. As the slide comes out of the water, it should be tilted so as to wrap the film

closely about it (Figs. 17 and 18). Care must be taken to ensure a tight fit of the film to the slide and to avoid pockets of air or of water above the section.

6. Hang the slide (wrapped in film), pegged at one end, on a rack to drain and to dry.

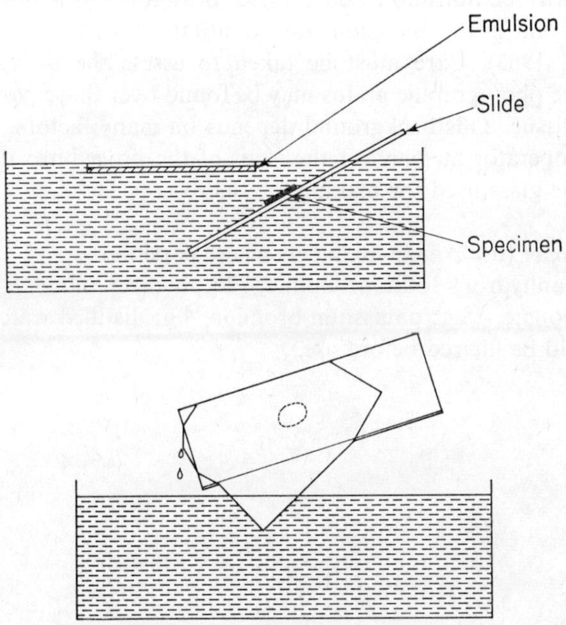

Figs. 17 and 18. The photographic film is floated on the surface of the water, emulsion side downwards. When it has expanded, a section mounted on a slide is immersed obliquely into the water, beneath the film. The slide is lifted out of the water and is tilted to wrap the film around the slide.

7. When dry, store in the dark in a light-tight dry container at 4°C or, for lowest background, at −25°C. The film is now exposed to the radiation. The time of exposure can vary greatly, depending on the radioactive matter used and the concentration which it has achieved in the tissue; it may be some days or some weeks.

8. The film is developed on the section, all the processing being done at 17–18°C to avoid damaging the film. It is essential to use freshly filtered D 19 b developer; Johnson's 'Fix-Sol' (1 : 9) is satisfactory for photographic fixation.

9. The section can then be stained with haematoxylin and eosin, or with methyl green-pyronin. These processes too should be done at 17–18°C.

Result. The photographic emulsion acts as a coverslip for the section but contains blackened regions or grains, caused by the radioactivity in the tissue section below it. Hence, at one focus, the tissue histology can be inspected, while at a slightly higher focus you should be able to observe how much radioactivity is contained in each tissue or cellular component. Quantitatively the number of photographic grains in the emulsion bears a definite (but not always exactly known) relationship to the number of radioactive atoms in the tissue below that point in the emulsion (for quantitation see Pelc, 1958; also Levi and Rogers, 1963). Care must be taken to assess the background count because some photographic grains may be found over those parts of the slide which lack tissue. This background depends on many factors, including the skill of the operator at many of the steps of the procedure; it may also be related to the glass used for making the microscope slide.

D 19 b developer (use Analar Reagents where possible)
Elon, 2·2 g; anhydrous sodium sulphite, 72 g; hydroquinone, 8 g; anhydrous sodium carbonate, 48 g; potassium bromide, 4 g; distilled water, 1000 ml.

This should be filtered before use.

PART VI: ENZYME HISTOCHEMISTRY

GENERAL INTRODUCTION

The study of enzymes by histochemical methods depends on a number of factors, as discussed below.

1. The physico-chemical characteristics of the enzyme itself. At present we have to use the data produced by conventional biochemistry. It must be realized that these data are derived, almost exclusively, either from isolated 'purified' enzymes or from enzymes in fractions separated from 'homogenized' cells. In a number of biochemical studies (reviewed by Siekevitz, 1962) it has been shown that the physical and chemical properties of several enzymes inside cell structures are markedly changed when the same enzymes are isolated from their structures. But for all that, until enzyme histochemistry becomes more precise (as discussed by Chayen and Bitensky, 1968), if we wish to discuss a particular enzyme, we must refer to those properties which biochemistry has found for it.

2. The properties of related enzyme systems. Very often it is difficult or even impossible to study one enzyme histochemically; one or more associated enzymes may be essentially involved in the histochemical demonstration of a particular enzyme. Thus, for example, it is rare (if ever) that histochemistry demonstrates lactate dehydrogenase alone; the enzyme is allowed to act on the lactate and in so doing, it reduces the co-enzyme (NAD→$NADH_2$). At least one other enzyme is required to transfer this hydrogen from $NADH_2$ to the tetrazole (or other hydrogen-acceptor). Hence it may be meaningless to argue about the precise histochemical localization of lactate dehydrogenase unless it is certain that both the lactate dehydrogenase and the hydrogen-transferring enzyme are located at the same site: the formazan is deposited by the latter, not the former enzyme. Similarly changes in the amount of oxidative activity (formazan deposition) shown for 'lactate dehydrogenase' in any given tissue at any given time may reflect changes in activity of the hydrogen-transferring enzyme rather than of the lactate dehydrogenase itself. But this can be tested by exposing a serial section to the reduced co-enzyme, $NADH_2$, and the tetrazole.

Because we are dealing with an histochemical system, various interactions can occur between the enzyme to be tested and other enzymes or reactive groups associated with that enzyme. For example, you might give a section glucose-6-phosphate and NADP, in an attempt to test how well it oxidizes this substrate (i.e. to test its glucose-6-phosphate dehydrogenase activity); it is hypothetically conceivable that the glucose-6-phosphate might be acted on so rapidly by a glucose-6-phosphatase in the section that little will be available for dehydrogenation. Provided one is aware of these possibilities, each can be investigated where the results indicate the need; but one must be aware of such differences between biochemical and histochemical analyses.

3. The properties of the subcellular environment. Generally, in biochemistry, the enzyme is first made fully available to its substrate. In contrast, the important contribution of histochemistry often lies in showing how available the enzyme is, rather than how much of the enzyme can be disclosed by 'suitable' (or 'unsuitable' in the physiological sense) disruptive techniques. In this way it becomes possible to begin to understand the control of cellular metabolism: the enzymes are present but are not working at full capacity all the time. The amount of activity manifested at any given time depends at least in part on the subcellular environment: on the permeability of membranes which may enclose the enzymes; on the state of the active groups which, in the intact cell, appear to be linked to the active site of the enzyme; or on the condition of associated enzyme arrays.

Hence in histochemistry it becomes pertinent to decide what you mean by 'a good reaction'. For too many histochemists this has come to mean maximal activity, preferably at a site which is believed to house the selected enzyme. It could be that the 'best' reaction (in the sense of meaning that which is most informative of the state of the enzyme in life) would be the least stained preparation. Consequently it may be advisable to obtain two results for each enzyme reaction: the first which shows how active the subcellular environment has allowed this enzyme to be in life (the manifest activity) and the second, after 'suitable' disruption of the subcellular organization, to show how much enzyme activity the tissue has potentially (the potential activity); the difference between the two indicates the latent, or reserve, activity held by the cell for emergencies.

4. The histochemical characteristics of the enzyme reaction. Histochemistry differs from biochemistry in that it attempts to precipitate the end-products of its reactions at the site at which they are produced; in this way it should be able to demonstrate the cellular or even the intracellular location of the enzyme, or enzyme complex, studied. But to localize the reaction involves precipitation, and for this phenomenon to occur, the end-product must be present in sufficient concentration as to be insoluble in the aqueous phase, i.e. in the incubation medium. These concepts require clear analysis.

We tend to speak of a substance, like calcium oxalate, as being 'insoluble' in water. This, of course, is imprecise; we mean that less than 0·000 67 g will dissolve in 100 parts of water (*Handbook of Chemistry and Physics*) but above this concentration it will precipitate. Calcium phosphate, under alkaline conditions, is even less soluble. But probably no salt is totally insoluble in water (e.g. see Adlam and Price, 1938, p. 131). When you have a saturated solution of a sparingly soluble salt, some is present in solution and since it is a salt, what little of it is dissolved is completely ionized. For all that, a sort of equilibrium is set up and this is described by the following:

$$\underset{\text{solid}}{Ca_3(PO_4)_2} \underset{\text{in equilibrium with}}{\rightleftharpoons} \underset{\text{dissolved}}{Ca_3(PO_4)_2} \rightleftharpoons \underset{\text{ionized}}{3Ca^{2+} + 2PO_4^{3-}} \qquad (1)$$

The equilibrium constant K is equal to the cube of the concentration of calcium ions multiplied by the square of the concentration of phosphate ions, and this is represented by $K = [Ca^{2+}]^3[PO_4^{3-}]^2$.

This constant, K, is the *solubility product*: if it is exceeded, then the equilibrium of equation (1) is disturbed and more calcium and phosphate ions must be precipitated as calcium phosphate (i.e. the reaction goes from right to left in equation (1)) to re-establish the value of K. This can also be represented as

$$K' = \frac{[Ca^{2+}]^3[PO_4^{3-}]^2}{[Ca_3(PO_4)_2]}$$

where the square brackets indicate 'concentration of' the substance within the brackets.

Thus, to keep K constant, if you add more phosphate ions (e.g. by enzymic cleavage of glycerophosphate) there must be a concomitant increase in calcium phosphate, i.e. you get precipitation of the end-product of the reaction.

Now we must consider a general case of the type that pertains to histochemical reactions. We have an enzyme which cleaves a substrate AB, and we wish to precipitate B. [*Note*: B can be the phosphate of glycerophosphate, or the hydrogen in a dehydrogenation process, or a naphthol derivative in the many coloured or azo-dye reactions to be described later.] We can precipitate B either by coupling it to a 'coupler' or by trapping it with a 'trapping agent', C, to yield an insoluble complex BC. The following factors have to be considered in analysing this process:

(*a*) *The solubility product of BC and enzyme activity*. As before, the solubility product is expressed by

$$K = [B][C] \text{ or } K' = \frac{[B][C]}{[BC]}$$

In order that K should be exceeded, so that BC should be precipitated immediately, *in situ*, the concentration of B and of C should be as high as possible. The concentration of the coupler or trapping agent (C) is under our own control in that we put a determined quantity of it into the incubation medium. [*Note*: the only problem is whether it is stable at the temperature and pH of the reaction, or whether a 'pulsing' procedure, as is needed for the aminopeptidase method, should be used to maintain a high concentration of the active form of the trapping agent.]

The less controllable factor is the concentration of B which is determined by the activity of the enzyme. If the tissue has been chemically fixed, the rate at which B is formed from AB can be greatly reduced: it is conceivable that, at any given moment, the concentration of B in a chemically fixed section might never be sufficiently high, at the site of a naturally weakly acting enzyme, to cause immediate precipitation of BC. Consequently, B would start to diffuse away from its site of origin and would set up a diffusion gradient. Whether enough of B would be formed ultimately to exceed the solubility product at the site of the enzyme would depend on the activity of the enzyme and on the rate of diffusion of B; the latter will be greatly affected by the nature of the diffusion gradient that is established (see below).

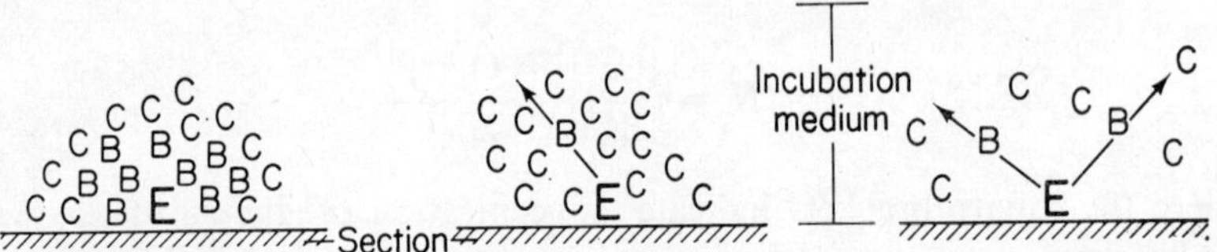

1. Active enzyme E liberates many molecules of B from AB and so exceeds the solubility product of BC. The high concentration of C in the incubation medium ensures that each molecule of B liberated is trapped by a molecule of C very close to the site of the enzyme.

2. This chemically fixed enzyme, E, shows only 10% of its unfixed activity. Hence in the same unit time, only one molecule of B is produced. Even though the concentration of C remains high, the solubility product of BC is not exceeded, and the molecule of B diffuses away before the next one is split off from AB by the enzyme.

3. In this case, the activity of E is weak but the concentration of C has been reduced by using an inadequate incubation medium. Consequently B does not encounter a molecule of C during its diffusion away from the site of its liberation (E).

(*b*) *Diffusion gradients and 'substantivity'*. When B is liberated from AB by the action of the enzyme, two antagonistic effects come into operation. The first is the tendency of B to become precipitated at the enzyme site; this is encouraged, histochemically, by having a high concentration of the trapping agent C. The second is the tendency of B to diffuse away from the site at which it is liberated. This is a widespread phenomenon in physical chemistry: each molecule or ion tries to fill the whole space available to it; the rate at which it does this depends on the difference in concentration of that molecule or ion at each point in the space. This can be understood more clearly by reference to simple examples: Suppose we have an enzyme acting on our substance AB in the absence of a trapping agent. You begin then with a high local concentration of B around the enzyme E, and some molecules of B will

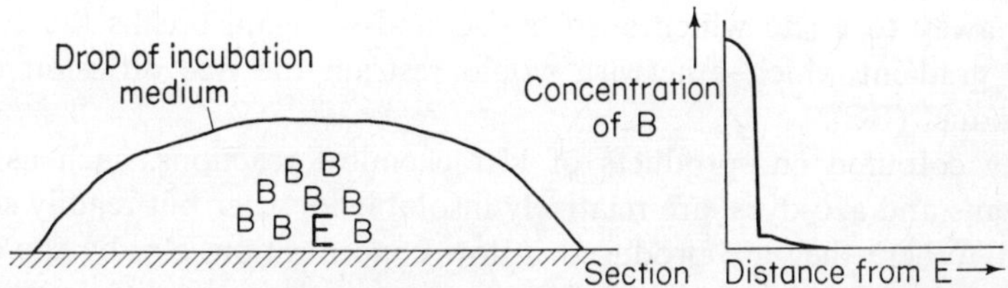

therefore begin to diffuse away from E to fill the incubation medium. Once this process has begun, a gradient of concentration of B will be established in the medium and this gradient will reduce the rate at which other molecules of B can move away from E.

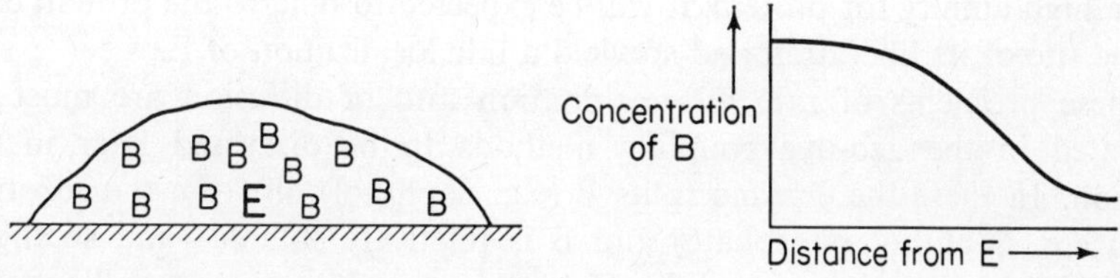

But suppose that, in the section and some distance from E, there is a site whose chemical composition or physical nature can bind B. Thus as each molecule of B reaches this site (S) it will be adsorbed on to it. In becoming adsorbed, B is no longer in solution at that point so that the gradient becomes broken at that point, i.e. the concentration of B *in solution*, becomes zero. This will provide the equivalent of a 'vacuum' and so it will actively draw B away from E to S. Provided that the binding site (S) is large enough, much or even all of B can be drawn away from E to S.

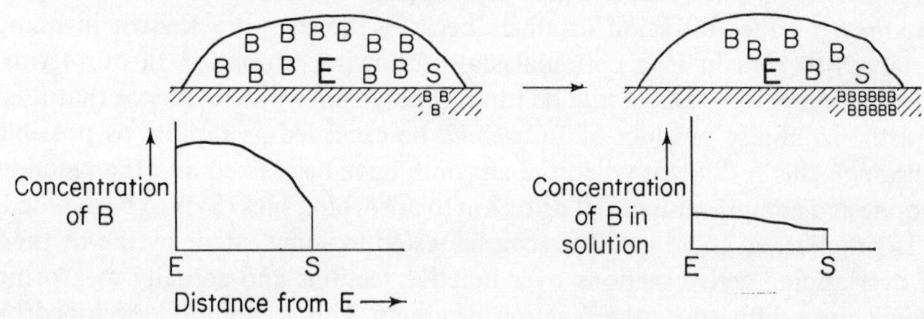

A similar situation will occur if insufficient trapping agent is added to the medium, so that B can diffuse without being captured adequately by the trapping agent. An analogous condition is one where weak enzyme activity (as occurs after certain types of fixation) produces too little of B to exceed the solubility product of BC at the site of the enzyme; the BC so formed can diffuse away to a site which adsorbs BC and so again breaks the concentration gradient which otherwise would restrain the free diffusion of the end-product (BC).

Many coloured end-products of histochemical reactions, such as some formazans and azo-dyes, are relatively insoluble in water but readily soluble in lipid. Where they are produced in low concentration, e.g. by the action of a weak (or of a chemically weakened) enzyme, they may diffuse into lipid and so set up a 'vacuum' in the diffusion gradient. For this reason many workers have been concerned over what they call the 'substantivity' of their end-products for lipids and for proteins. If the end-product has a high affinity ('substantivity') for lipid, you may expect this type of diffusion to cause a false localization of the dye at S instead of at E; if the end-product has a high affinity for protein, it will be expected to bind to the protein close to the site of its liberation and so yield a fair localization of E.

These problems of rate of precipitation and of diffusion are most aggravated in the azo-dye coupling methods to be discussed later in this Section. In these the enzyme splits B (e.g. naphthol) off from the substrate (AB; e.g. naphthol phosphate) and B is relatively soluble—and so highly diffusible—until it is coupled by C (e.g. hexazonium pararosaniline). The rate at which B and C couple depends on the pH, which is rarely optimal both for the enzyme activity and for the coupling. Consequently a compromise pH has to be used, which reduces the activity of the enzyme as well as the efficiency of the coupling reaction. The actual solubility of the end-product is not usually known; as Burstone (1962) explained: 'although there is no actual knowledge about the submicroscopic nature of a "capture reaction" involving a stain precursor (e.g. liberated naphthol), we can assume that this reaction should take place as rapidly as possible. A rapid capture reaction, however, does not necessarily preclude poor localizations. For example, β-naphthol may couple very rapidly with a suitable diazo reagent but may produce diffusion artefacts because the dyes are known in many instances to remain in a colloidal state following coupling.' In our terms, the reaction $B+C \rightarrow BC$ *should* be rapid (but we have little evidence that it is) and the solubility product of BC *should* be exceeded as rapidly as possible (although this is difficult when the enzymes have been fixed and the reaction is done at a compromise pH) if diffusion to adsorbing sites (S) is to be avoided.

In practice, all these variables can be tested to some extent by the method of overlapping active sections over inactive sections and actually measuring the distance diffused by the reaction products. This technique, developed by

Jacoby and Martin (1949) and particularly by Danielli (1953), has been used both by Seligman and his co-workers (Nachlas *et al.*, 1957a) and by McCabe and Chayen (1965) in attempts to analyse these effects.

CLASSIFICATION OF ENZYMES

Since 1961, enzymes have been systematized into an internationally agreed classification with rules governing the systematic and trivial names. A numerical system has also been developed by which the enzyme is characterized (for full discussion see Dixon and Webb, 1964). The new nomenclature has, as yet, had little impact on histochemistry so that in this book we retain the names in common usage in histochemistry although some reference will be made to the more precise nomenclature when discussing the biochemistry of enzymes. Data on the biochemical properties of each enzyme, listed below, have been taken mainly from Dixon and Webb (1964); *The Biochemists' Handbook* (1961; listed as *Biochemists' Handbook* where required); and from Roodyn (1965).

PHOSPHATASES

ALKALINE PHOSPHATASE

BIOCHEMICAL DATA

A range of alkaline phosphatases (E.C.* 3.1.3.1.) is known, their action being related apparently to the tissue and to the substrate which is preferentially (or more actively) used. They act on monoesters of orthophosphoric acid but have little effect on pyrophosphates, metaphosphates or on phosphoric diesters; intestinal alkaline phosphatase can hydrolyse amidophosphate bonds as occur in creatine phosphate (Dixon and Webb, 1964). The activity seems to depend on the presence of free —OH groups of tyrosine in the enzyme; it is believed that the enzyme functions by linking the phosphate to —OH groups of tyrosine or serine at its active site to form an intermediate 'enzyme-phosphate' (*Biochemists' Handbook*, 1961, p. 250).

Alkaline phosphatases are activated by metallic cations, particularly magnesium, and by certain amino-acids. As the name implies, they show optimal activity in the alkaline range from 7·6–9·9, depending on the concentration of the substrate and the activating ion. They can catalyse two types of reaction (*Biochemists' Handbook*, p. 249):

1. glycerol-1-phosphate + $H_2O \rightleftharpoons$ glycerol + inorganic phosphate,
2. glycerol-1-phosphate + glucose $\rightleftharpoons$ glycerol + glucose-6-phosphate.

They have been implicated in the maintenance of the intra-cellular concentration of phosphate for the formation of bone and in processes of absorption and transport across membranes (see *Biochemists' Handbook*; Danielli, 1953). In homogenates they are found mainly in the supernatant fraction although there is some suggestion that under certain conditions they can be found in the nuclear and microsomal fractions in rat liver (but see later under 'histochemical data' concerning diffusion of the enzyme).

HISTOCHEMICAL DATA

Alkaline phosphatase has been studied intensely by histochemists and many of the basic rules of enzyme histochemistry have been derived from studies on this enzyme (or group of enzymes). Reference should be made to Danielli (1953); or to Gomori (1952). Although there has been some discussion of

* E.C.: Enzyme Commission number nomenclature for specifying enzymes (Dixon and Webb, 1964).

how many phosphatases may occur in tissues, it seems clear that the term 'alkaline phosphatase' refers to a characteristic enzymic activity which splits phosphomonoesters optimally at an alkaline pH, i.e. greater activity is found at pH 9·4 than at 7·0. Consequently in a given tissue it is possible to show that there is optimal splitting of glycerophosphate at pH 9·4, and again at about pH 5 (corresponding to acid phosphatase) with some extra optimal activity (against nucleotide phosphate or against glucose-6-phosphate) at about pH 6·5. Even though some overlap of acid phosphatase, 5 nucleotidase and alkaline phosphatase activities can occur, the presence and characteristics of each can be distinguished.

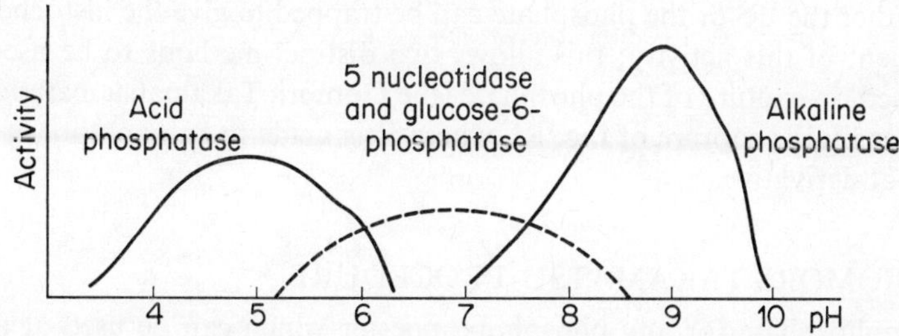

Alkaline phosphatase is one of the most popularly studied enzymes in histochemistry. Its practical advantage has been in aiding diagnosis in human liver biopsies (Sherlock, 1958; Bitensky, 1967a) and in the diagnosis and prognosis of leukaemia (see Wachstein, 1946; 1955; Hayhoe and Quaglino, 1958; Hayhoe, 1960). It occurs frequently at cell membranes where active transport occurs (see Danielli, 1953). It is found abundantly in the endothelium of blood vessels, in the brush border of renal convoluted tubules and at the lumenal margin of the intestinal epithelium.

Inhibitors. According to Barka and Anderson (1963), alkaline phosphatase can be inhibited if the sections are exposed to Lugol's iodine (2–3 min); to 15% acetic acid (2–3 min); or to 5% trichloroacetic acid (1–2 min). Burstone (1962) indicated that beryllium could be used as an inhibitor of this enzyme; this may be related to the effect found by Chèvremont and co-workers on the alkaline phosphatase of tissue culture cells which had been exposed to beryllium in life (Chèvremont and Firket, 1951a and b; 1953). Burstone (1962) listed cysteine and iodine at concentrations of 10^{-3}M, and sodium arsenate (10^{-2}M) as effective inhibitors; polyanions like polyoestradiol phosphate, polyphloroglucinol and polyphloretin were considered to be strong inhibitors of alkaline phosphatase of the kidney and intestine. Gomori (1953) regarded cysteine and cyanide as strong inhibitors of alkaline phosphatase.

The enzyme is inhibited to a greater or lesser degree by various chemical fixatives and by some embedding procedures (see Appendix).

Rationale of the histochemical methods

The reaction is

$$\underset{\text{phosphomonoester of R}}{\mathrm{R{-}O{-}\overset{\overset{\displaystyle O}{\|}}{\underset{\underset{\displaystyle OH}{|}}{P}}{-}OH}} + H_2O \rightleftharpoons ROH + \underset{\text{phosphoric acid}}{\mathrm{HO{-}\overset{\overset{\displaystyle O}{\|}}{\underset{\underset{\displaystyle OH}{|}}{P}}{-}OH}}$$

Thus either the 'R' or the phosphate can be trapped to give the histochemical assessment of this activity; this allows two distinct methods to be used, the one based on capture of the phosphate (the Gomori–Takamatsu method) and the other on the capture of the 'R', where R is made to be naphthol or some naphthol derivative.

THE GOMORI–TAKAMATSU PROCEDURE

Glycerophosphate (or any phosphomonoester which can be used at a high pH and a high concentration of calcium) is given to the section at pH 9·4 in the presence of activating magnesium ions. The phosphate which is liberated is trapped by the high concentration of calcium ions in the medium and precipitated as calcium phosphate which, *at alkaline pH values*, is virtually insoluble. The calcium phosphate is converted into the still insoluble cobalt phosphate and this is transformed into the black cobalt sulphide.

The main problems involved in this procedure are whether the enzyme, or the reaction products, may diffuse from their original site. Barter *et al.* (1955) proved, by interferometry, that once the calcium phosphate was precipitated, the subsequent procedures had no effect on its localization. A considerable controversy has raged, however, over whether nuclear localization of this enzyme activity was artefactual, due to diffusion of the enzyme, or of the phosphate (see Danielli, 1953; Gomori, 1952). It certainly appears, in fresh sections, that the enzyme itself can be lost from the sections into the incubation medium and may give rise to such an artefact, but that the enzyme is immobilized by a sufficient concentration of calcium ions in the incubation medium (Butcher and Chayen, 1966a). The concentration of calcium required for this immobilization did not appear to inhibit the activity of the enzyme, as has been suggested by some workers.

METHOD (after Gomori, 1952; Butcher and Chayen, 1966a)

Although many workers have recommended the use of frozen-dried sections we strongly recommend that unfixed cryostat sections should be used.

1. Incubate sections in freshly prepared medium at 37°C for up to 60 min (e.g. 8 μ sections of kidney may require 10 min while those of liver may need 60 min).
2. Wash in running tap water for 5 min.
3. Immerse in 2% cobalt nitrate solution for 5 min.
4. Rinse in distilled water.
5. Immerse in 0·5% (v/v) ammonium sulphide solution for 1 min.
6. Wash in distilled water.
7. Mount in Farrants' medium.

Result. Enzyme activity is shown by a black precipitate of cobalt sulphide.

Solutions required for this method

Incubation medium

2% barbitone sodium (sodium 5,5-diethylbarbiturate)	10 ml
3% sodium-β-glycerophosphate	10 ml
2% calcium chloride dried (or 2·7% $CaCl_2.2H_2O$)	20 ml
5% magnesium sulphate ($MgSO_4.7H_2O$)	1 ml
Distilled water	5 ml

Final pH of this medium is 9·4.

Control. Replace the 3% sodium-β-glycerophosphate with 10 ml of distilled water.

THE NAPHTHOL PHOSPHATE METHOD (Gomori, 1952)

In this procedure it is the 'R' (i.e. the naphthol) of the phosphomonoester which is trapped by simultaneous coupling with a diazonium salt. The reactions are shown on p. 110.

METHOD

Fixation is not recommended (see Gomori, 1952, p. 138 for discussion); it is advisable to use fresh unfixed cryostat sections which give results with this technique comparable to those obtained by the calcium phosphate method.

1. Incubate sections in freshly prepared incubation medium in a Coplin jar for up to 15 min.
2. Wash in distilled water.
3. Wash in 1% acetic acid for 1 min.
4. Rinse in distilled water.
5. Mount in Farrants' medium.

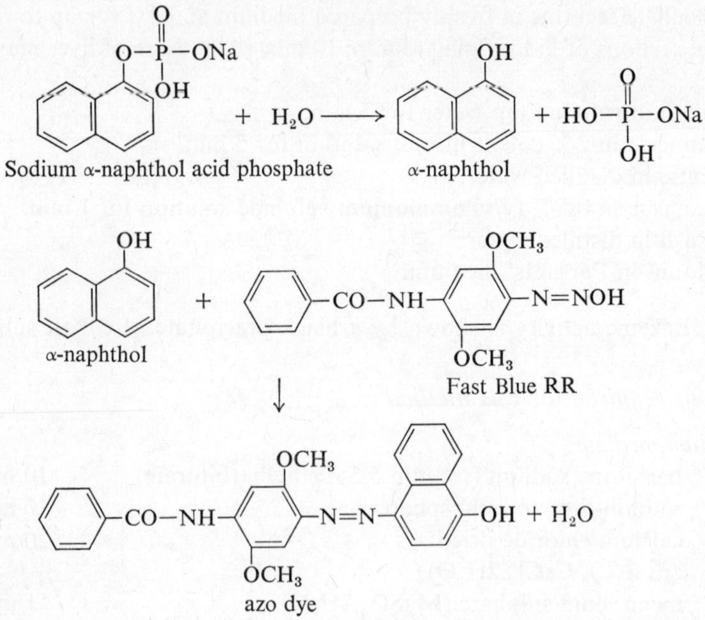

Sodium α-naphthol acid phosphate $+ H_2O \longrightarrow$ α-naphthol $+ HO{-}P(=O)(OH){-}ONa$

α-naphthol + Fast Blue RR ($-CO-NH-$, OCH_3, OCH_3, $-N{=}NOH$)

↓

azo dye ($-CO-NH-$, OCH_3, OCH_3, $-N{=}N-$, $-OH$) $+ H_2O$

Result. A purple-black precipitate denotes sites of alkaline phosphatase activity.

Solutions required for this method

Incubation medium

2% barbitone-sodium (sodium 5,5-diethylbarbiturate)	25 ml
α-naphthyl acid phosphate (sodium salt)	10 mg
10% magnesium chloride	0·2 ml
Fast Blue RR salt	25 mg

The final pH of the medium is 9·2.

Stir in the Fast Blue RR salt, filter into a Coplin jar and use IMMEDIATELY.

[*Note:* sections which were put into the incubation medium 20 min after it had been prepared were negative, assumably due to the decomposition of the Fast Blue RR salt.]

Control. As the test, but with no naphthyl acid phosphate in the incubation medium.

ACID PHOSPHATASE

BIOCHEMICAL DATA

From the full and detailed work of de Duve and his collaborators (e.g. see de Duve, 1957; 1959; Baudhuin *et al.*, 1964; Beaufay *et al.*, 1964) it now seems clear that the acid phosphatase (Enzyme Commission No. 3.1.3.2) of most tissues occurs inside lysosomes ('the light mitochondrial fraction'). A polar group close to the phosphate ester link can inhibit the action of this enzyme; complete inhibition has been found in one acid phosphatase when the phosphomonoester group and a free amino group (as in phosphoethanolamine or phosphoserine) are attached to adjacent carbon atoms; if they are separated by even one carbon atom there is no inhibition (Dixon and Webb, 1964). However the acid phosphatases from different animal tissues may behave differently (*Biochemists' Handbook*, pp. 250–51); their properties also seem to vary depending on the substrate given to the enzymes. Thus the pH optimum is usually between pH 4·5 and 5·5 but that of prostatic phosphatase, acting on phosphorylcholine, is at pH 6·5; this phosphatase catalyses the hydrolysis of most phosphomonoesters, of creatine phosphate and of amido-phosphate. The acid phosphatase of erythrocytes is activated by magnesium and is dependent on free —SH groups; these properties are not shared by other acid phosphatases.

Sodium fluoride seems to be a selective inhibitor of this enzyme in many tissues (Gomori, 1952).

The reaction sequence is the same as for alkaline phosphatase.

HISTOCHEMICAL DATA

The original method of Gomori (1952) was notoriously capricious. This is now known to be due to two major factors which were not known at that time. The first was the fact, demonstrated by Holt (1959) in his valuable cytochemical study, that the concentration of lead is critical and that the normal incubation mixture is unstable with respect to its content of lead. Holt's modified procedure is now widely used and has been the basis for much of the recent cytochemical work on this enzyme. The second is the fact that, in most tissues, this enzyme occurs inside lysosomes. These subcellular structures have peculiar properties: they keep their enzymes behind a relatively impermeable membrane so that little manifest activity can be demonstrated in intact tissue simply because the substrate generally used cannot get at the enzyme which is behind the membrane. But at the optimal pH for demonstrating the free enzyme (biochemically), the membrane itself becomes unstable so that more and more substrate can penetrate to the enzyme as the membrane progressively alters in its permeability. Fixation renders the membrane totally permeable.

The special value of the histochemical study of lysosomal enzymes (see Chayen and Bitensky, 1968) lies in two features: first it can localize the site of the enzymes and describe whether they are free or bound, and secondly it can disclose how much of the enzyme was active in life, in comparison with how much was retained in the latent state. This latter feature has proved of especial advantage in studying the effects of disease (Bitensky, 1963a); of immune reactions (Bitensky, 1963b, 1967b); and of cell damage (Chayen and Bitensky, 1968). Yet it must be emphasized that this advantage derives predominantly from the fact that the lysosomal membrane is retained in its relatively impermeable state; this excludes the use of freeze-drying or of chemical fixation.

Acid phosphatase occurs in most tissues. Its significance ranges from its effect during the cleavage of egg cells (see Dalcq, 1962, 1963); its role in disease, etc., as mentioned above; to its significance in plant cells (Gahan, 1967).

THE GOMORI LEAD PHOSPHATE METHOD (after Holt, 1959; Bitensky, 1962, 1963c)

Rationale. As with the alkaline phosphatase reaction, the sections are given glycerophosphate and the phosphate which is liberated has to be trapped. The reaction has to be done at an acid pH; under these conditions calcium phosphate is not insoluble. Consequently a different trapping agent has to be used and so lead is substituted for the calcium of the alkaline phosphatase method. The lead phosphate so formed is then converted directly into the brown-black lead sulphide by treatment with hydrogen sulphide (Bitensky, 1963c); this is preferable to ammonium sulphide because it avoids the drastic effects on the tissue of changing it from an acidic to an alkaline pH.

There have been many worries about spurious effects caused by 'metallophilia' (Barka and Anderson, 1963, p. 242), namely by the adsorption of the lead on to sites in tissues even in the absence of an added phosphomonoester. In fact, such enzymatically 'non-specific' adsorption of lead often gives valuable information concerning the chemical nature of the adsorbing site (see Ebel's test for polyphosphate, p. 68). It should not be a real worry provided that the specific reaction is controlled by the use of fluoride or of some other inhibitor.

Of greater concern is the fact that, in many tissues, a short reaction shows granular localization while prolonged reaction at the acidic pH produces intense diffuse and even nuclear staining. The probable explanation of this phenomenon is that the lead phosphate is not merely precipitated on or around the site of the acid phosphatase, but that it becomes adsorbed by some tissue component (possibly lipid in nature) at this site. With prolonged exposure to acidic conditions, this component (with its adsorbed lead phosphate) then diffuses about the cell and has a high affinity for the nucleus (Holt, 1963, p. 376). Consequently in unfixed tissues short incubation times show (*a*) where the enzyme is located; (*b*) the degree of manifest activity of this enzyme in life; (*c*) the fragility of the lysosomal membrane (Bitensky, 1963a). Prolonged incubation can indicate the total enzymic activity of which the tissue is capable but the localization is suspect. Controlled modification of the lysosomal membrane can allow a more precise measure of the latent activity of this enzyme.

METHOD

Fresh, unfixed cryostat sections should be used.

1. Incubate serial sections for various periods of time* in the reaction medium at 37°C in a Coplin jar.
2. Wash thoroughly in distilled water; it is essential to remove all the lead which is not precipitated or specifically adsorbed.

[*Note:* it is better not to use acetic acid for this wash because Desmet (1962) has shown that it removes the specific precipitate as well as the 'background'.]

3. Place the sections in distilled water which has been saturated with hydrogen sulphide. The time is not critical (10 sec to 1 min).
4. Wash thoroughly in distilled water.
5. Mount in Farrants' medium.

Control. Incubate as for the test, but include 0·01M sodium fluoride in the reaction medium.

Result. Black or brown-black deposit of lead sulphide. Only that reaction which is positive in the test but negative in the control, can be taken to be due to acid phosphatase activity.

Solutions required for this method

Reaction medium

To 400 ml of 0·05M acetate buffer at pH 5·0 (see Appendix) add 0·53 g of lead nitrate and 40 ml of a 3% solution of sodium β-glycerophosphate. In our experience the results vary slightly depending on the source of the β-glycerophosphate. That from B.D.H. gives clean results.

The whole reaction medium is left for 24 hr at 37°C during which time some precipitate forms. The solution must be allowed to cool before the solution is filtered. The reaction medium should be used immediately after it has been filtered.

* In rat liver 5–20 min incubation should suffice to give precise localization of a reasonable degree of activity; the exact time will depend on the age and strain of the rat and on physiological factors. In the human endometrium, 20 min incubation is required at mid-cycle; 40 min for the proliferative phase; but less than 5 min for the secretory phase. For total activity, at least twice the short incubation time should be given.

Solution of hydrogen sulphide

The hydrogen sulphide is made in a Kipps apparatus by the action of concentrated hydrochloric acid on zinc (or ferrous) sulphide. It should be bubbled through a wash bottle containing very dilute hydrochloric acid (to wash the gas) and the gas is then bubbled into a Coplin jar which contains distilled water. If 50 ml of distilled water are used, the gas should be passed into it for 10 min at a rate of about 120 bubbles a minute.

THE AZO-DYE ACID PHOSPHATASE METHOD (after Barka, 1960; Barka and Anderson, 1963)

Rationale. This method is comparable to the azo-dye technique for alkaline phosphatase. However, as with the Gomori calcium phosphate method, the acidic pH enforces certain changes. The efficient coupling of most diazonium salts requires an alkaline pH (Barka and Anderson, 1963, p. 242). Barka showed that if pararosaniline is diazotized just before it is used, the resultant hexazonium pararosaniline will couple with α-naphthol at pH 5·5 but its rate of hydrolysis (destruction) at this pH is greater than is its rate of coupling. Consequently a compromise is reached by doing the whole reaction at pH 6·5. There is remarkably little evidence concerning the speed of coupling at this pH; concerning how this pH affects the rate of enzymic liberation of the naphthol from the naphthol phosphate; or concerning the insolubility of the naphthol-diazotized pararosaniline complex, i.e. the rate at which it will precipitate. It is therefore difficult to evaluate this method with respect to the discussion given in the general introduction on p. 104. Barka and Anderson (1963, p. 244) agree that it is a disadvantage to do the reaction at pH 6·5 because at this pH the acid phosphatase is only half as active as it is at pH 4·5–5·0; moreover, at this pH, the hexazonium pararosaniline itself causes over 54% inhibition of the enzyme (so apparently leaving only a quarter of the activity of which the enzyme was capable of asserting after it had been fixed chemically).

It also appears, from our own results and from those reported by Holt (1963) and Novikoff (1963a) that the acid phosphatase demonstrated by this technique is markedly different from that characterized biochemically. It is therefore rather difficult to suggest its significance. However, the method is widely used and is included for this reason.

METHOD

Barka and Anderson (1963) recommend the use of frozen sections of tissue which has been fixed in cold formalin. We have obtained results with unfixed cryostat sections. The method given below is taken from Barka and Anderson (1963).

1. Mount the frozen sections on slides. Dry these at room temperature (1–2 hr).
2. Incubate at room temperature 5–30 min.
3. Rinse in water, dehydrate and mount in a synthetic resin.

Result. Red to red-brown indicates positive result. Fluoride does not seem to act significantly as a control for this reaction.

Solutions required for this method

Michaelis buffer

Sodium acetate.$3H_2O$	9·714 g
Sodium barbiturate	14·714 g
Distilled water to make up to	500 ml

Solution A: 4 mg/ml of sodium α-naphthyl acid phosphate in the Michaelis buffer.

Solution B: Add 2 g of pararosaniline hydrochloride to 50 ml of 2 N hydrochloric acid. Heat gently. Cool and immediately filter.

Solution C: 4% solution of sodium nitrite in distilled water.

Solution D: Diazotize the pararosaniline by adding 1 volume of Solution C to 1 volume of Solution B. Adjust the pH to 6·5 by adding sodium hydroxide (about 0·6 ml is recommended). Filter this solution.

Incubation medium

Add 5 ml of Solution A to 13 ml of distilled water; to this add 1·6 ml of the filtered Solution D.

[*Note:* Barka and Anderson warn that not all samples of pararosaniline are suitable; in some the impurities may be so excessive as to give 'non-specific background staining'.]

They also recommend the use of substituted naphthols, such as naphthol AS–TR which can be used with the hexazonium pararosaniline at pH 5·0. It is not clear how this overcomes the objections which Barka and Anderson (1963) have expressed to the use of this coupler at a low pH (as quoted above). However, the use of naphthol AS derivatives, including naphthol AS– and naphthol AS–BI phosphates, is widely recommended by many experts in this branch of enzyme histochemistry. From the work of Lojda *et al.* (1967) the final concentration of the naphthol AS derivative should be 0·5 mg per ml and that of the hexazotized pararosaniline should be 0·15 ml per ml of reaction medium (Lojda *et al.*, 1967).

5-NUCLEOTIDASE

BIOCHEMICAL DATA

5-Nucleotidase (Enzyme Commission No. 3.1.3.5) specifically catalyses the hydrolysis of phosphate esters on the carbon-5 of ribosenucleosides and of deoxyribonucleosides; it is inactive against phosphate esters on the carbon-3 of these molecules, and against phosphates of sugars, glycerol or phenols (*Biochemists' Handbook*, p. 255):

$$\text{adenosine 5-phosphate} + H_2O \longrightarrow \text{adenosine} + \text{phosphate.}$$

The pH optimum of this enzyme seems to vary considerably in different tissues (Hardonk and de Boer, 1968; also *Biochemists' Handbook*, p. 255). In the mouse kidney it shows a sharp optimum at about pH 7·5. But in mouse testis, brain and liver it has a broader activity : pH curve with optima at 8·5, 7·5–8·5 and about 8·0 respectively; the activities at pH 6·5 were at least 50% of that shown at the optimal pH. Hardonk and de Boer (1968) showed that the differences in these activities could be due to the occurrence of isoenzymes, some of which had acidic, some neutral and some alkaline pH optima. 5-Nucleotidase is activated by both magnesium and manganese ions but it is inhibited by fluoride and by borate (*Biochemists' Handbook*, p. 256); according to Dixon and Webb (1964) it may also be inhibited by calcium ions. A 5-nucleotidase, prepared from potatoes, is more than three times as active at pH 5 than at pH 9·4 but it is less specific in its activity at the lower pH (*Biochemists' Handbook*, p. 256).

Considerable interest has been shown in this enzyme in recent years because it seems to occur in very many different types of cell and to be a good marker for the plasma membrane (e.g. Emmelot and Bos, 1965; Fleischer *et al.*, 1969; Cheetham *et al.*, 1970). This enzyme activity may well be related to the adenyl cyclase (cyclic AMP) system (for details of this system see Robison *et al.*, 1970).

HISTOCHEMICAL DATA

Barka and Anderson (1963) advised rat liver or adrenal as the test tissue for this reaction; in the latter it was said to occur mainly in the medulla. Burstone (1962) showed high activity in the posterior lobe of the hypophysis. Hardonk (1968) found strong activity in many tissues; some tumours also showed strong reactions (Hardonk and Koudstaal, 1968). According to Burstone (1962) manganese ions are stronger activators than are magnesium

ions (also see Hardonk, 1968); nickel (Ni^{2+} at 10^{-3}M) and zinc almost completely inhibit 5-nucleotidase, leaving the non-specific alkaline phosphatase relatively unaffected; 0·1M sodium fluoride and 0·08M sodium borate are said to inhibit the 5-nucleotidase reaction completely (Heppel and Hilmoe, 1951).

Rationale of the histochemical procedures

Basically this is a simple phosphatase reaction in which the liberated phosphate has to be trapped as in the Gomori methods for acid and alkaline phosphatases (pp. 108 and 112). There are two practical problems: the first is that this reaction must be specific for the phosphate ester on the carbon-5 of the nucleoside and the second is that the non-specific alkaline phosphatase will also hydrolyse such phosphates with such intensity that if both 5-nucleotidase and non-specific alkaline phosphatase are present at the same locus in the tissue, it is difficult to distinguish between them since both are demonstrated by the deposition of the phosphate liberated equally by the specific and the non-specific enzyme. For this reason the reaction for 5-nucleotidase is done at a lower pH, at which the activity of the non-specific alkaline phosphatase is considerably diminished. For example, at pH 8·3 the 5-nucleotidases with alkaline pH optima are optimally active but the activity of alkaline phosphatase may be reduced to one-third of its maximal activity (Gomori, 1952, p. 186). Alternatively the 5-nucleotidase reaction can be done at an even more acidic pH (6·5 or 7·0). These are optimal for the neutral and acidic 5-nucleotidases. At this pH calcium phosphate is not insoluble so lead has to be used as the trapping agent; this avoids the use of calcium which has been claimed to cause some inhibition of the isolated enzyme (but there is no evidence that lead may not be equally inhibitory). However, if the lead method is used, the reaction must be differentiated from that caused by the acid phosphatases of the tissue, which are as non-specific as are the alkaline phosphatases.

But to some extent these cautionary notes emphasize the value, not the limitations, of histochemistry. In this type of study we are less concerned with the precise physical chemistry of the enzymes than with how the tissue deals with a particular substrate: (*i*) can it hydrolyse this type of substrate; (*ii*) does it hydrolyse it specifically, as it might if this reaction has some particular part to play in the metabolism of this tissue; (*iii*) does this reaction occur in some particular site in the cell: this might indicate the significance of this reaction in a particular cell function, e.g. acid phosphatase in lysosomes and 5-nucleotidase in the plasma membrane. As histochemists we might then be concerned with the physical chemistry of the hydrolysis of defined phosphate esters at specific intracellular sites, irrespective of which of a number of enzymes, at those sites, were involved.

Thus, depending on the pH at which the reaction is to be performed, there are two methods which can be used for the histochemical demonstration of 5-nucleotidase, depending on whether the liberated phosphate is trapped with lead or with calcium. The rationale of these is dealt with under the acid and the alkaline phosphatases respectively (see pp. 108 and 112).

LEAD METHOD FOR 5-NUCLEOTIDASE (after Naidoo and Pratt, 1954; Wachstein and Meisel, 1952, 1957)

METHOD

Use unfixed, fresh, cryostat sections.

1. Incubate sections at 37°C in freshly prepared medium in a Coplin jar for up to 20 min.
2. Wash in running tap water.
3. Rinse in distilled water.
4. Immerse in a 0·5% (v/v) solution of ammonium sulphide* for 1 min.
5. Rinse in distilled water.
6. Mount in Farrants' medium.

Result. Sites of enzyme activity are shown by a brown precipitate of lead sulphide.

Solutions required for this method

Incubation medium

Adenosine-5′-monophosphate	31 mg
0·1M acetate buffer pH 6·5	40 ml
0·1M lead nitrate	1 ml

Final concentration of AMP is 2×10^{-3}M.

Control. Replace the AMP with 25 mg (2×10^{-3}M) sodium-β-glycerophosphate.

CALCIUM METHOD FOR 5-NUCLEOTIDASE (after Gomori, 1952, p. 186)

METHOD

Use fresh, unfixed cryostat sections.

1. Incubate sections in freshly prepared medium at 37°C for up to 20 min.
2. Wash in running tap water for 5 min.
3. Immerse in 2% cobalt nitrate solution for 5 min.
4. Rinse in distilled water.
5. Immerse in 0·5% (v/v) ammonium sulphide solution* for 1 min.

* Ammonium sulphide may be replaced by H_2S-water, as in the acid phosphatase method, or by a 2% solution of sodium sulphide, pH 8·0.

6. Wash in distilled water.
7. Mount in Farrants' medium.

Result. Enzyme activity is shown by a black precipitate of cobalt sulphide.

Solutions required for this method

Incubation medium

Adenosine-5′-monophosphate	31 mg
0·2M Tris buffer, pH 8·3	20 ml
2% calcium chloride solution	20 ml
10% magnesium sulphate solution	0·5 ml

Final concentration of substrate is 2×10^{-3}M.

Control. Replace the AMP with 25 mg (2×10^{-3}M) sodium-β-glycerophosphate.

GLUCOSE-6-PHOSPHATASE

BIOCHEMICAL DATA

This enzyme (Enzyme Commission No. 3.1.3.9) is tightly bound to the microsomes of mammalian liver, kidney and intestinal mucosa (de Duve *et al.*, 1955; for many studies see Roodyn, 1965); it can be released as a lipo-protein molecule by treatment with deoxycholate (Beaufay and de Duve, 1954). It is found in highest concentration in these tissues but traces have been found in the spleen, heart, skeletal muscle, brain, lung and gastric mucosa of rats (Dixon and Webb, 1964). Its apparent function (*Biochemists' Handbook*, pp. 253–54) is to allow glucose to be released into the circulation. Its activity in the liver has been reported to be increased in animals given large doses of glucocorticosteroids; it is elevated in diabetic rats but can be restored to normal levels by injection of insulin; its activity is reduced in hypophysectomized rats. One of the hepatic glycogen-storage diseases is said to be due specifically to loss of this enzyme (see *Biochemists' Handbook*).

Glucose-6-phosphatase is most active at pH 6; when isolated it is most stable at pH 8 and is denatured at pH 5; it does not act on glucose-1-phosphate (*Biochemists' Handbook*).

HISTOCHEMICAL DATA

According to Chiquoine (1953, 1955) this enzyme activity is completely inhibited by fixation in 80% ethanol and embedding in paraffin wax (in contrast to alkaline phosphatase); it is also completely inhibited by fixation in cold formalin. Fluoride, zinc, cyanide and alloxan also inhibit it.

Rationale of the histochemical method

This depends on the ability of the enzyme in the section to split phosphate from glucose-6-phosphate specifically at a pH of 6·5. At this pH, the phosphate must be trapped by lead as discussed for acid phosphatase (p. 112); the precipitated lead phosphate is visualized by converting it into the coloured lead sulphide.

METHOD (after Wachstein and Meisel, 1956)

Fresh, unfixed cryostat sections should be used.

1. Incubate sections at 37°C in freshly prepared incubation medium in a Coplin jar for up to 20 min.

2. Wash well in running tap water.
3. Rinse in distilled water.
4. Immerse in 0·5% (v/v) ammonium sulphide solution for 1 min.
5. Wash in distilled water.
6. Mount in Farrants' medium.

Result. Sites of glucose-6-phosphatase activity are denoted by a brown precipitate of lead sulphide.

Solutions required for this method

Incubation medium

0·1M acetate buffer, pH 6·5	40 ml
Glucose-6-phosphate—Na salt	26 mg
0·1M lead nitrate	1 ml

Final concentration of glucose-6-phosphate is 2×10^{-3}M.

Control. Replace the glucose-6-phosphate with 25 mg (2×10^{-3}M) sodium-β-glycerophosphate to test the chemical specificity of the reaction.

ADENOSINE TRIPHOSPHATASES

(*a*) CALCIUM-ACTIVATED (MYOSIN-TYPE) ATPase

BIOCHEMICAL DATA

Although these ATPase enzymes are usually classed with the phosphatases they are not esterases; they are better classed with the pyrophosphatases in that they act on the phosphate-to-phosphate link instead of the phosphate-to-carbon link as in glycerophosphatases or nucleotidases:

$$A{-}P\sim P\sim P + H_2O \rightarrow A{-}P\sim P + H_3PO_4 + \text{energy}$$

adenosine triphosphate $\sim$P denoting the 'energy-rich' phosphate bond	adenosine diphosphate	phosphoric acid

In rat liver two types of ATPase are found; their localization in the different cell fractions is not all that dissimilar (Dixon and Webb, 1964):

	Nuclei	Mitochondria	Microsomes	Supernatant
Calcium-activated (myosin-type) ATPase (3.6.1.3)	+	++	+	−
Magnesium-activated ATPase (3.6.1.4)	+	+++	+	−

The enzyme which will be considered first is typical of myosin and in fact seems to be an inherent attribute of the myosin molecule (*Biochemists' Handbook*); it occurs in the rat in the liver; kidney (in high concentration); heart (in very high concentration); skeletal muscle (high activity); lung (high activity); pancreas and brain (Dixon and Webb, 1964). Hence, unless one is dealing with myosin itself, it is better to refer to it as the 'calcium-activated ATPase' (Enzyme Commission No. 3.6.1.3) rather than as 'myosin ATPase'. It has a molecular weight of 230,000 when isolated from dog heart, and 540,000 when purified from rabbit muscle (Dixon and Webb, 1964). It is stimulated by dinitrophenol (DNP), apparently because DNP and actin change the configuration of the protein (Dixon and Webb, 1964). It is inhibited when more than half its —SH groups are combined with *p*-chloromercuribenzoate although its activity may be even enhanced if less —SH groups are so combined. This enzyme is activated by calcium ions but inhibited by magnesium ions. There seems to be some doubt as to the biochemical pH optimum for this enzyme when 'purified' (see Burstone, 1962).

HISTOCHEMICAL DATA

In view of the fact that there may be three types of ATPase (one being activated by calcium and the other two by magnesium but having different pH optima) it is not surprising that the histochemical literature has been somewhat confusing. As with the true phosphatases, the histochemical demonstration of any of these enzymes depends on the splitting of a phosphate group from the substrate, at the correct pH and with the required activators, etc., and its capture by either lead or calcium, depending on the pH (as discussed for acid and alkaline phosphatases on pp. 108 and 112).

According to Barka and Anderson (1963), formalin fixation causes 80–90% inhibition of these enzymes and certain histochemical procedures cause an additional 85% inhibition, so leaving about 5% of the original activity. These strictures do not apply to the use of frozen sections and the calcium method, as developed by Padykula and Herman (1955a and b) in their excellent studies on myosin ATPase activity (in muscle). In agreement with the biochemical data these workers showed that the ATPase activity of the myosin was inhibited by *p*-chloromercuribenzoate ($2{\cdot}5 \times 10^{-3}$M) and could be re-activated by L-cysteine ($2{\cdot}5 \times 10^{-3}$M) or by BAL (2,3-dimercapto-1-propanol; concentration: 5×10^{-3}M). Moreover, by the use of their type of histochemical reaction, Niles *et al.* (1964b) were able to show that the ATPase activity of myosin was indeed activated by calcium ions and by DNP, but apparently inhibited by magnesium ions, in marked contrast to studies by other workers who used frozen-dried tissues and were unable to demonstrate these biochemical characteristics of myosin ATPase activity (see discussion in Burstone, 1962).

Doubts arose over the original method of Padykula and Herman (1955a and b) because the ATPase activity was not concentrated in discrete bands in striated muscle. This was shown by Niles *et al.* (1964b) to be due partly to poor preservation of the muscle in their frozen sections; partly to insufficient calcium ions in the incubation medium which were essential for the adequate capture of the liberated phosphate and for activation of the enzyme. Niles *et al.* (1964b) showed that precise localization of the myosin ATPase activity within the myofibril, at sites known to contain myosin, depended on the rapid release of the phosphate (which was enhanced by adding dinitrophenol to the medium) and on adequate capture of the released phosphate by a sufficiently high concentration of calcium in the medium. They showed that the activity was high at pH 9·4 but that there was no ATPase activity at pH 7·0 or at pH 5·5.

This histochemical work has been of considerable value in assessing the biochemical lesion in human myocardial dysfunction associated with prolonged open-heart surgery (Niles *et al.*, 1964a); its significance has been emphasized by Pearse (1964).

The method used is basically the calcium-trapping technique as discussed for alkaline phosphatase.

METHOD FOR CALCIUM-ACTIVATED ATPase ACTIVITY

Fresh cryostat sections, prepared as described in this book, must be used.

1. Incubate in the freshly prepared reaction medium at 37°C for 10–20 min.
2. Wash in three changes of 1% calcium chloride, 2 min in each bath.
   [*Note:* the presence of calcium ions in the water used for washing keeps the precipitate from dissolving, as discussed in the General Introduction to this Part.]
3. Transfer to 1% cobalt nitrate solution for 2 min and then sequentially into two baths of this solution, each for 2 min.
   [*Note:* this procedure changes the calcium phosphate into cobalt phosphate.]
4. Wash well in distilled water (to remove adsorbed cobalt).
5. Place in dilute ammonium sulphide (0·5% v/v) for 1 min.
6. Wash well in distilled water.
7. Mount in Farrants' medium.

Result. ATPase activity is shown by the brown-black precipitate of cobalt sulphide. It should act as a stain for the cross-striations of cardiac and voluntary muscle.

[*Note:* The precipitation of the phosphate at the site of its liberation depends on an adequate concentration of calcium. If the tissue is deficient in calcium, it may be advisable to pre-incubate the section for 5 min at 37°C in a 1% solution of calcium chloride before incubating the section with the substrate (step 1 above).]

Control. In muscle, it is simple to control this reaction by incubating a serial section under the same conditions but with glycerophosphate in the the place of the adenosine triphosphate. Otherwise it may be necessary to add $2{\cdot}5 \times 10^{-3}$M *p*-chloromercuribenzoate to the reaction medium.

Reaction medium (prepare immediately before use)

0·1M sodium barbiturate solution (2·06 g/100 ml)	10 ml
ATP	76 mg
2,4-Dinitrophenol	30 mg
Calcium chloride solution ($10{\cdot}5$ g $CaCl_2.2H_2O$/100 ml)	5 ml
Distilled water	35 ml

(*b*) MAGNESIUM-ACTIVATED ATPase

BIOCHEMICAL DATA

A magnesium-activated ATPase has been obtained from skeletal muscle and from insect muscle. The former was also slightly activated by manganese ions and was inhibited by calcium, by fluoride and by *p*-chloromercuribenzoate; it had optimal activity at pH 6·8–7·0. It appears to contain phospholipid which is essential for its activity (*Biochemists' Handbook*, pp. 257–58).

A magnesium-activated ATPase is found in mitochondria. Its pH optimum is said to be around pH 8·5 but it is markedly affected by the relative concentration of magnesium and of ATP. It is inhibited by fluoride and by *p*-chloromercuribenzoate (*Biochemists' Handbook*). This enzyme seems to be that designated by the Enzyme Commission No. 3.6.1.4, which is activated by magnesium and inhibited by calcium (Dixon and Webb, 1964), although there is evidence (Lehninger, 1965, p. 94) that there may be two or even three mitochondrial ATPase enzymes which can be distinguished by their different pH dependence. Intact mitochondria, in contrast to disrupted mitochondrial fragments, show no ATPase activity; dinitrophenol activates this (or these) enzymes both in intact mitochondria and when they have been rendered soluble (Lehninger, 1965).

The biochemical enzyme 3.6.1.4 has been studied in homogenate fractions by many workers (see Roodyn, 1965). It has been found in highest concentration in the mitochondrial fraction, with some spread to the nuclear and the microsomal fractions, the proportion found in these other fractions varying somewhat with different homogenization and fractionation procedures. It must be remembered that biochemists have diffusion and preparatory artefacts too; they are not confined to histochemistry.

Of some importance, in many aspects of cellular physiology and biochemistry, is the sodium and potassium-activated adenosine triphosphatase (Na^+–K^+-activated ATPase) of the 'sodium pump'. This enzyme system has been reviewed by Charnock and Opit (1968) and Skou (1965). It is an integral part of cell membranes of many if not all cell types. Its hydrolysis of ATP requires both Na^+ and K^+, although the latter appears to be replaceable by other ions; the hydrolysis is inhibited by calcium, fluoride and by other agents known to reduce active transport in cells. It is inhibited by cardiac glycosides, such as ouabain (10^{-4}M). Its pH optimum seems to be around pH 7·2–7·6. There seems good reason for believing that it is concerned with the transport of Na^+ and K^+ across cell membranes. The biochemistry of these ATPase enzymes is certainly not simple and some of the difficulties met with by histochemists (e.g. Tormey, 1966) may be due to the complexity of the system rather than to inadequacies of histochemical technique.

HISTOCHEMICAL DATA

Because this ATPase requires magnesium for its activation, it is inadvisable to use calcium as the trapping agent: magnesium phosphate is relatively soluble so that the magnesium not only activates the enzyme but also competes with the calcium for the phosphate (as shown in practice by Niles *et al.*, 1964b). For this reason, and because of the lower pH at which these enzymes are studied, lead has to be used as the trapping agent for the phosphate, liberated by the ATPase, even though it seems likely that lead may inhibit certain forms of ATPase activity to a greater or lesser extent (e.g. Pratt, 1954; Tormey, 1966; Jacobsen and Jørgensen, 1969). Dalcq (1962; 1963; Dalcq and Pasteels, 1963) has used a form of the Wachstein and Meisel (1956) procedure to valuable effect in his studies on cleavage in eggs of mammals and of invertebrates. Wachstein *et al.* (1960) used a modification of their own lead technique for the study of mitochondrial ATPase activity.

The rationale of the method is as for the acid phosphatase or 5-nucleotidase methods (p. 112).

There has been considerable controversy over the histochemical demonstration of ATPase activity by the lead method. Rosenthal *et al.* (1970) summarized their findings as follows: when the lead concentration is at least 1mM it induces the non-enzymatic hydrolysis of ATP, provided that the concentration of ATP is not more than 1mM. The free phosphate so liberated can give rise to reaction product artifacts (e.g. Rosenthal *et al.*, 1969b). The precise rate of non-enzymatic, lead-catalysed hydrolysis of ATP depends on the relative concentrations of lead and of ATP and on the presence of certain divalent ions (Tice, 1969). The reaction product is not lead phosphate but is a complex of lead, nucleotide and phosphate (Rosenthal *et al.*, 1969a). On the other hand, they agreed that this lead-catalysed artifactual hydrolysis of ATP cannot account for all the results obtained by this method (Ganote *et al.*, 1969).

The demonstration by Marchesi and Palade (1967) of a Na^{+}–K^{+}-stimulated ATPase, which was inhibited by ouabain, has been much discussed because of the low concentration of lead which was necessary for this histochemical reaction in order not to inhibit the enzyme and not to give lead hydrolysis of the ATP (see Novikoff, 1970a and b; Rosenthal *et al.*, 1970; Jacobsen and Jørgensen, 1969). Some workers considered that, at such low concentrations of lead, the efficiency of trapping would be so reduced that diffusion artifacts were likely to occur (Rosenthal *et al.*, 1969b; see Moses and Rosenthal, 1968, for fuller details). Although Novikoff (1970b) agreed that we still need more knowledge about the conditions under which lead complexes are precipitated in such cytochemical reactions, he found the lead method reasonably satisfactory (see Novikoff, 1970a; also Novikoff *et al.*, 1958; Novikoff, 1967).

The relative concentrations of lead and of ATP seem to be critical and can be modified by the ionic concentration of the incubation medium. Jacobsen and Jørgensen (1969) made a valuable contribution to this study. Although they were unable to demonstrate a Na^+–K^+-activated ATPase, their results led them to believe that the lead method could be used to locate the ATPase (or ATPases) which is probably involved in the transport of cations. They used 2mM lead, which produces barely detectable amounts of non-enzymic hydrolysis of ATP even in the presence of 3mM magnesium, in the presence of 3mM ATP (pH 7·3; 80mM Tris–maleate buffer). Concentrations of sodium and of potassium should be about 120mM and 30mM respectively (also see Tormey, 1966).

LEAD METHOD FOR ATPase (Wachstein and Meisel, 1956; Wachstein *et al.*, 1960)

METHOD

Fresh sections should be used.

1. Incubate in medium for 2 hr at 37°C.
2. Wash in running tap water.
3. Rinse in distilled water.
4. Immerse in 0·5% (v/v) ammonium sulphide solution for 1 min.
5. Wash in distilled water.
6. Mount in Farrants' medium.

Result. Sites of ATPase activity are indicated by a brown precipitate of lead sulphide.

Solutions required for this method

Incubation medium

ATP	25 mg in 20 ml distilled water
0·2M Tris buffer, pH 7·2	20 ml
2% lead nitrate	3 ml
0·25% magnesium sulphate	5 ml
Distilled water	2 ml

To this medium can be added 40 mg of DNP.

PHOSPHAMIDASE

BIOCHEMICAL DATA

There is relatively little known about enzymes which split phosphoamide bonds; often they are included in the acid phosphatases (as in the *Biochemists' Handbook*). Dixon and Webb (1964, p. 620) point out that energy is stored in two forms, either as pyrophosphate bonds (as in ATP) or as phospho-amide bonds in creatine or arginine phosphates.

$$HN{=}C\begin{cases}N(CH_3)CH_2COOH \\ N(H){-}P(O)(OH)OH\end{cases} \xrightarrow{+H_2O} HN{=}C\begin{cases}N(CH_3)CH_2COOH \\ NH_2\end{cases} + HO{-}P(O)(OH){-}OH$$

phosphocreatine
(the phospho-amide bond
is marked by the arrow)

creatine

phosphoric acid

The enzyme which acts in this way on phosphocreatine (Enzyme Commission No. 3.9.1.1) is found apparently fairly widely in animal and plant tissues and in snake venom (see Dixon and Webb, 1964, p. 761).

An additional point of interest is that it has been claimed by Holter and Li (1950, 1951) that crystalline pepsin, trypsin, chymotrypsin and rennin can cleave phospho-amide bonds; the optimal pH of phosphamidase activity in homogenates was around pH 4·6 (see Burstone, 1962). Such activity does not seem to fit in with the Enzyme Commission numbering but would be of great value in histochemistry, because phosphamidase activity might be used to demonstrate proteolytic activity generally.

HISTOCHEMICAL DATA

Gomori developed a histochemical method for phosphamidase in which the phospho-amide bond was that of *p*-chloroanilidophosphonic acid; he obtained results which could be interpreted clearly as being due to a specific phosphamidase enzyme only between pH 5–6. Others have used phosphocreatine at an alkaline pH but their results do not seem to be conclusive (see Gomori, 1952, p. 195). The particular interest in phosphamidase activity has been that it has frequently been found in high concentration in tumours and that there is evidence that the degree of malignancy is paralleled, more or less, by the concentration of this enzyme (see Gomori 1952, p. 196; also Burstone, 1962).

Gomori (1952) did not find his own method 'entirely dependable'. Burstone (1962) pointed out that the phospho-amide bond is labile so that some decomposition occurs during the incubation and that, even if the section is held at an inclined angle in a Coplin jar, you get an 'indiscriminate precipitation of lead phosphate over the tissues'. Other histochemical data are reviewed by Meyer and Weinman (1953, 1955) and by Barka and Anderson (1963).

This indiscriminate precipitate, and the variability of the method, have been overcome by Butcher's (unpublished) modification of Gomori's method. In this case the polyvinyl alcohol is used as a stabilizer of the substrate rather than primarily to protect the tissue. Unfixed sections are used; this is advantageous:

(*i*) because fixed tissue, having more free reactive groups as a result of the chemical denaturation, adsorbs lead more readily than does unfixed tissue;

(*ii*) because the enzyme is more active and so liberates phosphate more rapidly; consequently the solubility product (see p. 101) of lead sulphide is exceeded quickly, to yield a precipitate with less likelihood of diffusion;

(*iii*) because this enzyme appears to occur in discrete cytoplasmic organelles and to have the same sort of properties as lysosomal acid phosphatase; thus prolonged incubation at an acidic pH, or disruption of the particles by chemical fixation, will contribute to a confused histochemical result, just as with acid phosphatase. It is distinguished from lysosomal acid phosphatase by the fact that it is not inhibited by fluoride. Moreover its activity in invasive tumours, as indicated by Gomori (1952), and its location at the lumenal margin of the small intestine (Butcher, unpublished data) indicate that it is a separate, and physiologically significant enzyme.

METHOD (after Gomori, 1952; Meyer and Weinman, 1955; Butcher, unpublished data)

Use fresh cryostat sections for this reaction.

1. Place section in incubation medium at 37°C in a Coplin jar for up to 20 min.
2. Wash sections well in running tap water.
3. Rinse in distilled water.
4. Immerse in 0·5% (v/v) ammonium sulphide solution for 1 min.
5. Wash in distilled water.
6. Mount in Farrants' medium.

Result. Sites of phosphamidase activity are denoted by a brown precipitate of lead sulphide.

Solutions required for this method

Incubation medium

Solution A

0·05M acetate buffer, pH 5·4	50 ml
0·1M (3·3 g/100 ml) lead nitrate	1·9 ml
Sodium chloride	0·11 g
Polyvinyl alcohol (MO5/140)	5 g

Add the lead nitrate and sodium chloride to the buffer. Then dissolve the PVA by heating, stirring continuously. Cool to 37°C.

Solution B (this must be prepared immediately before it is to be used)

p-Chloroanilidophosphonic acid	0·15 g
1M sodium hydroxide	0·5 ml

Add solution B to solution A in a Coplin jar.

Control. (*i*) Omit solution B. Any 'reaction' will be due to adsorption of lead.

(*ii*) Add 0·01M of sodium fluoride (21 mg/50 ml) to the incubation medium. True phosphamidase activity will not be inhibited.

ESTERASES

BIOCHEMICAL DATA

It is necessary to consider something of the physical chemistry of the esterases if the behaviour and classification of these enzymes is to be understood even superficially. Careful work has been done on carboxylesterase by Webb (see Dixon and Webb, 1964, pp. 218–20) and reference will be made to this work in our review of the elements of the enzymology of esterases. Much of what will be described has application to other enzymes.

ELEMENTARY ENZYMOLOGY

An organic ester is the result of reaction between an organic acid and an alcohol (analogous, superficially, to the reaction of an acid and a base to yield an inorganic salt). A simple example is the reaction between acetic acid ($CH_3.COOH$) and ethyl alcohol (C_2H_5OH) to yield ethyl acetate ($CH_3.COOC_2H_5$):

$$CH_3.COOH + H.O.C_2H_5 \rightarrow \underbrace{CH_3.CO}_{\text{acetyl group}}.O.\underbrace{C_2H_5}_{\text{alkyl group}} + H_2O$$

The ester bond is represented by

$$-\overset{\overset{\large O}{\|}}{C}-O-\overset{|}{\underset{|}{C}}-$$

Similarly esters can be formed between 'inorganic' acids and alcohols as for example, between phosphoric acid and ethyl alcohol:

$$3H-\overset{H}{\underset{H}{C}}-\overset{H}{\underset{H}{C}}-O\boxed{H+HO}-\overset{\overset{O}{\|}}{\underset{OH}{P}}-OH \rightarrow H-\overset{H}{\underset{H}{C}}-\overset{H}{\underset{H}{C}}-O-\overset{\overset{O}{\|}}{\underset{\underset{C_2H_5}{O}}{P}}-O-C_2H_5 + 3H_2O$$

It will be seen that in this case the ester bond is denoted by

$$-\overset{|}{\underset{|}{C}}-O-\overset{\overset{O}{\|}}{\underset{|}{P}}-$$

whereas in the ethyl acetate it was $-\overset{|}{\underset{|}{C}}-O-\overset{\overset{O}{\|}}{C}-$. Consequently it will be seen that the phosphatases and esterases of histochemistry are both esterases which cleave somewhat similar ester bonds. The specificity depends on the substitution of phosphorus for carbon on one side of the bond. A phosphatase, therefore, is a specific type of esterase (e.g. a phosphomono-esterase).

The straight-chain esters hydrolysed by carboxylesterase can be written, in general formula, as acyl$-\overset{\overset{O}{\|}}{C}-O-$alkyl (where acyl is the general form of acetyl in the specific acetic-ester, above). They become aligned at specific sites on the enzyme molecule (Fig. 19).

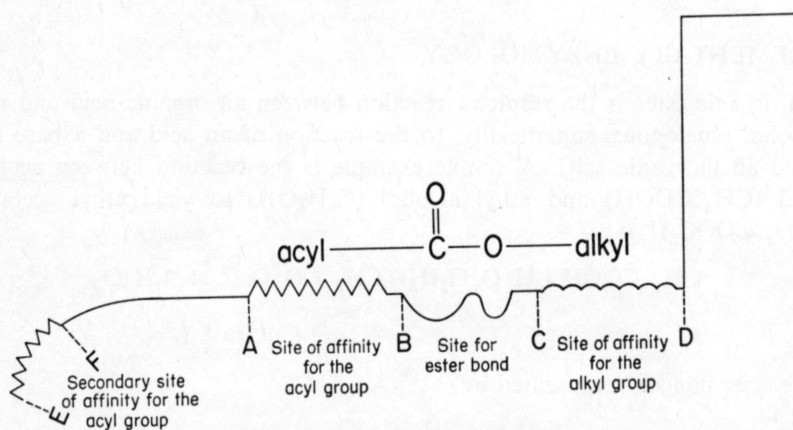

Fig. 19. The orientation of the substrate on the active surface of the esterase.

Webb showed that the affinity and the reactivity of the ester was increased if the acyl group was extended up to a chain length of 4–6 carbon atoms; further increase in chain length gave increased affinity, by 'adhesion' at the secondary site (E—F) but this changed the orientation at B—C so that there was decreased reaction. A similar effect was found for the alkyl group: up to 4–6 carbon atoms gave greater affinity and reactivity but increase in chain length above this reduced both affinity and reactivity; this indicates the size of C—D and the fact that there was no secondary site for the alkyl group as there was for the acyl group. The region B—C binds the ester bond and so locates the substrate molecule on the face of the enzyme molecule.

Since this enzyme acts on both aliphatic (straight-chain) and on aromatic (benzene-ring) esters, this work of Webb (see Dixon and Webb, 1964) has relevance also to 'naphthol-acetate esterase' and to the lipases; in the latter the different chain lengths of the acyl group which can be bound on the enzyme depends assumably on the different lengths of A—B and of E—F in the different lipases.

INHIBITION OF ESTERASE ACTIVITY (see Dixon and Webb, 1964, pp. 346 *et seq.*)

Many enzymes which have esterase activity are inhibited in a highly specific manner by organo-phosphorus substances which have the general formula

$$(R{-}O)(R{-}O)P(=O)X \quad \text{and} \quad (R{-}O)(R')P(=O)X$$

where R and R′ are alkyl groups and X can be $-CN$, $-F$ or $-O.C_6H_4.NO_2$. These substances also inhibit trypsin, chymotrypsin, plasmin and thrombin. Their mode of action depends on the way esterases function. The substrate is first aligned on the esterase, as in Fig. 19. The acyl group (Ac) is then transferred to the enzyme (Enz.), so forming an acylated enzyme and liberating the alkyl group (Alk) as the free alcohol (1). The acylated enzyme then becomes hydrolysed (2):

$$Ac{-}C({=}O){-}O{-}Alk + Enz.H \rightarrow Ac{-}C({=}O){-}Enz. + HO.Alk \qquad (1)$$

$$Ac{-}C({=}O){-}Enz. + H_2O \rightarrow Ac{-}C({=}O){-}OH + Enz.H \qquad (2)$$

The organo-phosphorus substance mimics this process. Thus it too becomes aligned on the enzyme; the enzyme becomes phosphorylated and the X group is liberated (as was the Alk group of the normal ester). The difference

$$(R{-}O)(R')P({=}O)X + Enz.H \longrightarrow (R{-}O)(R')P({=}O)Enz. + HX \qquad (3)$$

between the behaviour of the normal ester and the organo-phosphorus substance now becomes apparent because the rate at which the phosphorylated enzyme can be hydrolysed is much slower than the rate at which the acylated enzyme is hydrolysed (2). In the weaker inhibitors this hydrolysis (4) does occur at a measurable rate and the enzyme can then become fully active again; in the more permanent inhibitors the hydrolysis occurs too slowly to have physiological value. This very slow hydrolysis is the basis of the methods

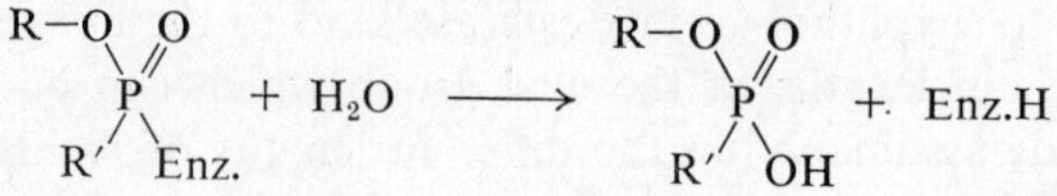

$$\begin{array}{c}R{-}O\quad O\\ \diagdown\ /\!/\\ P\\ /\ \diagdown\\ R'\quad Enz.\end{array} + H_2O \longrightarrow \begin{array}{c}R{-}O\quad O\\ \diagdown\ /\!/\\ P\\ /\ \diagdown\\ R'\quad OH\end{array} + Enz.H \qquad (4)$$

of Ostrowski and his colleagues (Ostrowski and Barnard, 1961; Ostrowski *et al.*, 1963) for the histochemical localization of acetylcholinesterase, in which the enzyme inhibitor is labelled radioactively, so that the enzyme is located, autoradiographically, by the site of the radioactive inhibitor.

The action of eserine and of prostigmine, which are powerful inhibitors of both cholinesterase and of acetylcholinesterase, is due to their acting very much as do the organo-phosphorus substances (Dixon and Webb, 1964, p. 355, quoting work of Wilson *et al.*, 1960). They become aligned on the enzyme at the active esterase site (B—C in Fig. 19) and their carbamoyl group is transferred to the enzyme to give a carbamoyl-enzyme which is only very slowly hydrolysed (the reaction being that of (2) and (4)). The enzyme finally does recover by its slow decarbamoylation by water (hydrolysis); the speed of this decarbamoylation can be considerably enhanced by hydroxylamine.

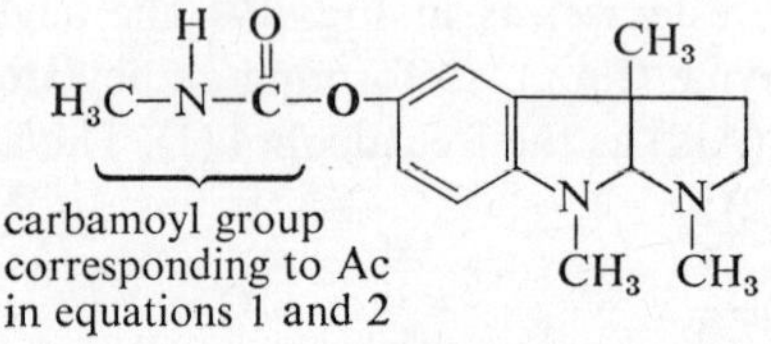

H O
$H_3C{-}N{-}C{-}O{-}$ [ring structure] CH_3
N N
CH_3 CH_3

carbamoyl group corresponding to Ac in equations 1 and 2

Eserine
The heavier type indicates the ester bond

NOMENCLATURE AND PROPERTIES

The biochemistry of the esterases is very difficult and this is reflected in the very confused nomenclature and literature on the whole subject (see Dixon and Webb, 1964, p. 221). At one time the term 'ali-esterase' was used to denote an enzyme which attacked only aliphatic esters; later evidence (as discussed above) indicates that this is not a valid concept. Similarly lipases are considered as different from other esterases only in the nature of regions E–F and A–B (Fig. 19) and in that they act on substrates which generally are not in true solution.

The main classification of esterases depends on the distinction between those (B-type esterases) that are inhibited by a low concentration of organo-phosphorus compounds (i.e. those that perform reaction (4) at a very slow or negligible rate) and those that are not (A-type esterases which do perform reaction (4) reasonably rapidly). Cholinesterases are particular examples of B-type esterases in that they are also inhibited by low concentrations of eserine. The classification is reviewed by Aldridge (1956, 1961).

The cholinesterases are also of two types. The first, which used to be called 'true cholinesterase' but is better referred to as acetylcholinesterase (Enzyme Commission No. 3.1.1.7), is actually inhibited by high concentrations of acetyl choline. The second type, which used to be called 'pseudo-cholinesterase', is now referred to as cholinesterase (Enzyme Commission No. 3.1.1.8). Neither type of cholinesterase requires the charged nitrogen atom of choline for its activity although cholinesterase is more specific for the choline structure than is acetylcholinesterase (Dixon and Webb, 1964). Both types hydrolyse aromatic esters (like naphthol acetate used in histochemistry). The real difference between the two types seems to be that the rate of hydrolysis by acetylcholinesterase decreases sharply as the chain length of the acyl group is increased whereas the reverse is true of cholinesterase (Dixon and Webb, 1964).

In tissues acetylcholinesterase (3.1.1.7) has been found in rat liver, occurring in greatest concentration in the microsomal fraction; similarly carboxylesterase (3.1.1.1) has been found in greatest concentration in the microsomal fraction of the liver of the mouse and rat (Roodyn, 1965).

Some of the chemical characteristics are listed in Table 5.

HISTOCHEMICAL DATA

In view of the confusion over the biochemistry of isolated esterases, there seems little point generally in trying to be pedantic over esterase activity found in tissues. Anyone wishing to be so is referred to the thoughtful reviews in books by Gomori (1952), Burstone (1962), Barka and Anderson (1963). It seems more reasonable to use histochemistry to demonstrate esterase 'activity' and to try to characterize that activity by rigorous histochemical analysis (e.g. with various substrates and inhibitors).

Histochemistry has been of particular value in the study of cytoplasmic esterase activity in that it has shown this activity to be located in discrete particles which resemble lysosomes. This is in sharp contrast to the biochemical localization of this activity in the microsomal fraction. This problem was investigated by a combination of biochemical and histochemical procedures in a collaborative study by Underhay *et al.* (1956); recent biochemical studies have confirmed the lysosomal location (Tappel, 1969).

Histochemical esterase activity has been reported to be high in the liver, pancreas, 'chief cells' of the gastric mucosa, kidney tubules and bronchial mucosa (Burstone, 1962). There have been promising reports of significant esterase activity in tumours (e.g. see Gomori, 1955) and in a very careful study Willighagen *et al.* (1963) have suggested that the histochemical reaction for this type of activity may be of prognostic value.

Cholinesterase, or acetylcholinesterase (by the Koelle method) has been much and successfully used as a stain for nervous tissue (see, e.g., Garrett, 1966a and b).

In more precise histochemical studies, various inhibitors have been used in attempts to be more definite about the type of esterase activity which has been demonstrated. The present evidence indicates that considerable caution should be exercised in interpreting the results of this type of study. One of

TABLE 5
Some properties of esterases

	Acetylcholinesterase (*'true cholinesterase'* 3.1.1.7)	*Cholinesterase* (*'pseudo-cholinesterase'* 3.1.1.8)	*Lipases* (*glycerol ester hydrolase* 3.1.1.3)	*A-esterases*	*B-esterases*
Effect of eserine ($1–5 \times 10^{-5}$M)	Inhibited	Inhibited	Unaffected	Unaffected	Unaffected
Effect of organophosphorus compounds (usually 10^{-5}M)	Inhibited	Inhibited	Some lipases are unaffected	Unaffected (these compounds are readily hydrolysed)	Inhibited
Occurrence	Nerve-tissue erythrocytes; liver	Serum; heart; intestinal mucosa; pancreas	Pancreas; oats		
pH optima	7·5–8·5	7·5–8·5	8·0 (but 6–7 in presence of taurocholate)	7·8	
Activators	NaCl; KCl; $CaCl_2$		Ca and Mg	Mn; glyoxaline; histidine; metal-chelating agents	

the commonly used inhibitors is 'ISO-OMPA' (tetraisopropylpyrophosphophoramide; Burstone, 1962); according to Barka and Anderson (1963), when this is used at a concentration of 3×10^{-5}M it causes 90% inhibition of cholinesterase, whereas the compound 62C–47 completely inhibits acetylcholinesterase (at concentrations of $3 \times 10^{-5}–1 \times 10^{-4}$M). They comment, however, that the inhibitory effect of these, and other, compounds varies in different organs even in the same species; different species of animals respond in different ways.

NAPHTHOL ACETATE METHOD FOR ESTERASE

Rationale. The ester used is that formed from acetic acid (acyl group) and α-naphthol (in place of ethyl alcohol on p. 131). As has been discussed (above) esterases will act on aromatic or on aliphatic esters; the advantage of having naphthol as the alkyl group is that as it is liberated, it can be coupled to form an insoluble azo-dye (as in the acid phosphatase method, see p. 114). The method is simple and much used. One note of warning must be added. In the discussion of the biochemistry of esterases we noted that these enzymes act on the specific ester bond. We also noted the similarity between the $-\overset{|}{\underset{|}{C}}-O-\overset{\overset{O}{\|}}{C}-$ and the $-\overset{|}{\underset{|}{C}}-O-\overset{\overset{O}{\|}}{\underset{|}{P}}-$ bonds. The basis of inhibition by organo-phosphorus compounds (see above) was their ability to locate themselves at the active site of the esterase molecule; they inhibited only because their rate of hydrolysis off the active site was so slow. Thus it seems surprising that many histochemists apply naphthol acetate to one section, and naphthol phosphate to another and claim, without further controls, that the former shows esterase activity while the latter shows phosphatase activity. With such staining procedures, some control, and some caution in interpretation, might be helpful.

METHOD (after Gomori, 1952)

Use fresh, unfixed cryostat sections.

1. Incubate sections at room temperature for up to 10 min in the reaction medium.
2. Wash in several changes of distilled water.
3. Mount in Farrants' medium.

Result. Enzyme activity is shown by a purple colour.

Reaction medium

Dissolve 10 mg of α-naphthol acetate in 0·25 ml of acetone. Add this to 20 ml of 0·1M phosphate buffer at pH 7·4. Shake well. Then dissolve 20 mg of Fast Blue B salt in this solution. Filter the solution on to the section; use *immediately*.

Control. The control used will depend on what type of esterase activity is being investigated (see biochemical and histochemical data, above).

INDOXYL ACETATE METHOD FOR ESTERASE

Rationale. Barrnett and Seligman (1951) suggested that indoxyl acetate could be used as an ester substrate; esterase would liberate the indoxyl group which would become oxidized to indigo. Holt (1952, 1958; Holt and O'Sullivan, 1958; Holt and Sadler, 1958; Cotson and Holt, 1958; Holt and Withers, 1958) developed this idea; there are few cytochemical methods which have been studied with such physical–chemical precision. Basically the reaction sequence is as follows (Holt and Withers, 1958):

O–C(=O)–CH$_3$ (indoxyl acetate, positions 1–7, 2CH, 3C, N–H) I —esterase→ OH (indoxyl, C, CH, N–H) II

oxidation by potassium ferricyanide–potassium ferrocyanide → (indigo: C=O, C=C, N–H) III

Of many indoxyl acetates tested, Holt and his collaborators found that the 5-bromo-4-chloroindoxyl acetate gave the most precise staining. The concentration of the oxidant was stressed by Cotson and Holt (1958) because the step from II to III goes via a *leuco*-dye and they suggested that if the rate of oxidation is too slow (e.g. if aerial oxidation is relied upon) then the *leuco*-indigo may diffuse to artefactual sites before it becomes coloured. The rate of oxidation is depressed at lower pH values (Cotson and Holt, 1958); the effect of the oxidant may be somewhat variable under a number of conditions.

There has been some criticism of the method as given originally by Holt, on the grounds that the formalin which he used as a fixative completely inhibits one form of esterase and that the ferri- ferro-cyanide oxidant is also inhibitory (see Shnitka and Seligman, 1960, 1961; Seligman, 1963).

METHOD

Unfixed cryostat sections should be used.

1. Incubate in the reaction medium at 37°C for 15–30 min.
2. Wash in 30% alcohol containing 0·1% acetic acid.
3. Wash in distilled water.
4. Mount in Farrants' medium.

Result. Esterase activity is shown by a blue colour.

Solutions required for this method

Reaction medium

Tris buffer (0·2M), pH 8·5	2 ml
Oxidant	1 ml
1M calcium chloride	0·1 ml
2M sodium chloride	5·0 ml
Distilled water	2·0 ml

Add this solution rapidly, with agitation, to 1 ml of absolute ethyl alcohol containing 1·3 mg of 5-bromoindoxyl acetate.

The oxidant is prepared as follows:

Potassium ferricyanide	210 mg
Potassium ferrocyanide	155 mg
Distilled water to make up to	100 ml

Control. As discussed for the naphthol acetate method.

LIPASE

Rationale. The substrate for lipases is a Tween: these compounds are esters of long-chain fatty acids and either sorbitan or mannitan (Gomori, 1952). The enzyme splits the fatty acid off and this is trapped as the insoluble calcium salt. The latter is then converted into the lead salt in order to produce a dark sulphide which stains the tissue. There is some doubt as to whether the rate of capture of the fatty acids by the calcium is sufficiently rapid, and whether the calcium soaps so produced are sufficiently insoluble (in both water and lipid) as to provide rigorous intra-cellular localization of the site of the reaction. However, clear reactions have been obtained, which relate to specific parts of tissues known to contain these enzymes, and different responses have been observed to Tweens which have fatty acids of various chain length. We are indebted to Dr. F. A. Fairweather for his modification of Gomori's technique; he has used it in his detailed studies of the small intestine in idiopathic steatorrhoea. A different, potentially valuable, method is that of Seligman *et al.* (1966).

METHOD (Fairweather's modification of the procedure of Gomori, 1952).

Fresh cryostat sections should be used.

1. Incubate in the reaction medium at 37°C for 2–3 hr.
2. Wash well in distilled water, preferably containing 1% calcium chloride to ensure that the calcium soaps remain insoluble.
3. Place in a 1% solution of lead nitrate for 15 min.
4. Wash in running tap water for 5 min.
5. Immerse for 1 min in water saturated with hydrogen sulphide (as for the Gomori acid phosphatase method, p. 114).
6. Wash well in tap water.
7. Mount in glycerine jelly or in Farrants' medium.

Result. Brown-black precipitate of lead sulphide indicates lipase activity.

Control. As for the test but leave the Tween out of the reaction medium (any colour is then due to 'metallophilia').

Solutions required for this method

Stock solutions (*to be prepared fresh on the day of use*)

A. 5% aqueous solution of the selected Tween (e.g. Tween 60).
B. 0·5M Tris buffer, pH 7·2.
C. 10% solution of calcium chloride.

Reaction medium:

Tris buffer (B)	5 ml
Tween solution (A)	2 ml
Calcium chloride solution (C)	2 ml
Distilled water	40 ml

CHOLINESTERASE

Rationale. Acetylthiocholine has been shown to be hydrolysed by both acetylcholinesterase and by cholinesterase at either the same rate, or even more rapidly than acetylcholine (see Gomori, 1952, p. 210). Moreover, as discussed in the section on the biochemical data of esterases (above), longer chain esters, such as butyrylthiocholine, are acted on by cholinesterase more rapidly than by acetylcholinesterase and this distinction has been used in histochemistry (see Gomori, 1952; Barka and Anderson, 1963). The thiocholine liberated by such enzymatic activity is precipitated by reaction with copper, present as copper glycinate; the copper thiocholine is then transformed into brown copper sulphide by treatment with ammonium sulphide. Traces of copper thiocholine are added to the various fluids used in this reaction in order to ensure that any formed by the enzymic reaction will remain insoluble (i.e. you add enough of the copper thiocholine to ensure that the solubility equilibrium, discussed on p. 101 is kept from right to left in the equation; thus there is less danger of the enzymatically formed copper thiocholine being brought into solution to maintain the solubility constant K'). For a discussion of the chemistry of this reaction, see Malmgren and Sylvén (1955).

Useful variants of this technique are the gold-thiocholine and gold-thiolacetic acid methods for acetylcholinesterase and for cholinesterase (Koelle and Gromadzki, 1966).

METHOD (after Koelle and Friedenwald, 1949)

Fresh cryostat sections should be used (Barka and Anderson (1963) emphasized that formalin fixation results 'in considerable loss of enzyme activity').

1. Incubate in the reaction medium at 37°C for 10–60 min, depending on the tissue. [It is convenient if the sections are held horizontally so that the freshly prepared reaction medium can be filtered directly on to them.]
2. Wash in distilled water containing traces of copper thiocholine.
   [*Note:* this compound should be almost insoluble so only enough should be added to provide a true solution of it. There should not be so much as to form a suspension which may become adsorbed on to the section.]
3. Dip into ammonium sulphide solution (as for alkaline phosphatase) and leave for 1 min.
4. Wash in distilled water.
5. Mount in Farrants' medium.

Result. Brown deposit (often crystalline) indicates cholinesterase or acetylcholinesterase activity.

Controls (see Garrett, 1966a)

(*i*) 3×10^{-5}M eserine should inhibit both cholinesterase and acetylcholinesterase activity. It can be added to the normal incubation medium.

(*ii*) ISO-OMPA at a concentration of 3×10^{-6}M should inhibit cholinesterase activity. It may be necessary to treat sections for 30 min with this, in the reaction medium which lacks acetylthiocholine, before incubating in the full reaction medium (+ISO-OMPA+acetylthiocholine). Any reaction after this procedure may be considered to be due to *acetyl*cholinesterase.

(*iii*) B.W.284C51 or 62C–47, at concentrations of 3×10^{-5} to 1×10^{-4}M are expected to inhibit acetylcholinesterase. As with ISO-OMPA, it may be necessary to pre-incubate for 30 min in the reaction medium containing the selected inhibitor but lacking the substrate, and then to incubate in the full reaction medium which contains both the substrate and the inhibitor. For this it is better to use butyrylthiocholine as the substrate. Any reaction after this procedure may be interpreted as being due to cholinesterase activity.

Solutions required for this method

Copper thiocholine

Dissolve acetylthiocholine iodide in a 1% solution of copper glycinate (to give an excess of copper glycinate). Leave this solution to stand overnight. Collect the precipitate (which should be copper thiocholine); wash it with water several times.

Solution A: Dissolve 3·75 g of glycine in 18 ml of 1 N potassium hydroxide. Make up to 100 ml with distilled water.

Solution B: 0·1M solution of cupric sulphate.

Solution C: Dissolve 14·5 mg of acetylthiocholine iodide in 0·75 ml of distilled water. Add 0·25 ml of solution B. Cupric iodide is precipitated leaving the acetylthiocholine in solution. Centrifuge and decant, retaining the clear solution.

Reaction medium

Distilled water	8·6 ml
Solution B	0·2 ml
Solution A	0·4 ml

To this add a trace of copper thiocholine (ideally this should be added to make the reaction medium 2×10^{-3}M with respect to copper thiocholine).

Heat to 37°C.

Just before use, add 0·8 ml of solution C.

Filter on to the section (i.e. it must be used immediately).

PROTEASES

BIOCHEMICAL DATA

At one time it was thought that there were two types of proteolytic enzymes, the proteinases which attacked large molecules and degraded them to smaller molecules which were then acted on by the peptidases. It now seems that the critical factor is the chemical group present on either side of the peptide bond. Certain peptidases require to have a terminal carboxyl or amino group on one side of the peptide bond (these are the exopeptidases); others (the endopeptidases) may even be inhibited by such bonds but act more efficiently if a particular amino-acid is present. For example, pepsin has been said to attack the peptide bond between the α-carboxyl group of a dicarboxylic acid (the other carboxyl group being free) and the α-amino group of an aromatic amino acid; it is inhibited if a free amino group is present in the vicinity. Trypsin attacks the peptide bond produced by the carboxyl group of arginine and of lysine; its activity depends on the second amino group of these residues remaining free and it can be inhibited by the close proximity of carboxyl radicals. (For fuller discussion see Bergmann, 1942; Bergmann and Fruton, 1941; Dixon and Webb, 1964, pp. 225 *et seq.*). Such enzymes obviously preferentially act on bonds inside the protein chain (hence the name 'endopeptidase'); because the end-products cannot be precipitated for histochemical study, the only possibility of demonstrating them histochemically comes either from detecting associated activity (as, for example, with phosphamidase activity, see p. 128) or by allowing the enzymes to leak out into a protein film and to demonstrate the areas of the film which become affected (see Daoust, 1965, for a review of these methods for deoxyribonuclease, ribonuclease, amylase and proteases).

One other biochemical point deserves to be stressed. Unlike pepsin, certain proteases, particularly trypsin (Dixon and Webb, 1964, p. 243) split amides (cf. phospho-amidase) and esters even more rapidly than they hydrolyse peptide bonds. Even though trypsin is a powerful endopeptidase, it rapidly hydrolyses lysine ethyl ester. Chymotrypsin also hydrolyses amides and esters and shows preference for aromatic esters (Dixon and Webb, 1964, pp. 243–45); it may therefore be expected to have some activity against the substrate used for the histochemical 'aminopeptidase' reaction. Cathepsin C also acts on amides and esters and can be listed as an 'aminopeptidase' (see Dixon and Webb, 1964, p. 246). Leucine aminopeptidase requires a free α-amino group and so is a true exopeptidase or aminopeptidase. Unlike trypsin, it will not act on esters of amino acids (although it will hydrolyse amides) and, correspondingly, it is not inhibited by di-isopropyl phosphorofluoridate. (Organo-phosphorus compounds inhibit enzymes which have esterase activity; see p. 133). It is not specific for leucine.

It may be relevant to note that apart from the proteolytic enzymes discussed above, plasmin or fibrinolysin acts as an endopeptidase (Burstone, 1962, p. 395).

Relevance to histochemistry

It will be realized, from these biochemical data, that proteolytic enzymes can appear, histochemically, through their esterase activity, and there has been some discussion as to whether the apparently lysosomal localization of indoxyl acetate esterase activity (in view of the biochemical evidence of the non-lysosomal location of 'esterase') could be due to proteolytic activity. It is also apparent that some proteolytic enzymes will affect 'amidase' reactions, such as the phosphamidase reaction (p. 128). Finally, in view of this biochemical evidence, there seems little reason for believing that the histochemical 'aminopeptidase' reaction should not be given by a whole range of proteolytic enzymes. For these reasons, the phosphamidase and the aminopeptidase reactions may prove more valuable than they would if they demonstrated only their nominal restricted activity.

AMINOPEPTIDASES AND NAPHTHYLAMIDASES

BIOCHEMICAL DATA

Many enzymes seem to have aminopeptidase activity. Much of the biochemical work on 'aminopeptidase' has been done by Smith and collaborators (see Smith, 1951, 1960; Smith and Spackman, 1955). Renal aminopeptidase seems to occur within a particulate fraction; to have a pH optimum of 7·0–8·5, depending on the substrate, and to be activated by cobalt (*Biochemists' Handbook*, p. 295). On the other hand, 'amidase' (Enzyme Commission No. 3.5.1.4) which acts on leucine amide, is found predominantly in the supernatant fraction (Roodyn, 1965, p. 164; *Biochemists' Handbook*, p. 286); it has optimal activity at about pH 8.

Nachlas *et al.* (1957a, 1962) first developed the use of leucine-β-naphthylamide (also called leucine 2-naphthylamide) for investigating aminopeptidase activity histochemically and this chromogenic substrate has been much used biochemically. Patterson *et al.* (1963, 1965) showed that aminopeptidases, such as hog kidney aminopeptidase, hydrolyse this substrate about 15,000 times more slowly than the more conventional leucineamide, but that there were other enzymes in their homogenates which appeared to act on leucine 2-naphthylamide more rapidly than they did on leucineamide. It is now known (e.g. Marks *et al.*, 1968; Behal *et al.*, 1969) that there is a distinct group of enzymes which have little or no leucineamide activity (i.e. these are not aminopeptidases) but which rapidly hydrolyse amino acid-β-naphthylamides. These enzymes are now classified as arylamidases or naphthylamidases. Their physiological substrate is not yet known although it has been suggested that their function may be to inactivate regulatory peptides such as angiotensin II amide (Behal *et al.*, 1969) or to liberate bradykinin and to activate neurosecretory material (Marks *et al.*, 1968). Hopsu-Havu *et al.* (1966) showed that an aminopeptidase could remove the terminal lysine from the 10-peptide kallidin-10 and leave it as the stable 9-peptide, bradykinin. The stability of bradykinin is apparently due to the fact that its terminal amino acid is proline, which does not seem readily hydrolysed from the rest of the peptide.

The arylamidases, or naphthylamidases, have been much studied. They occur in highest concentration in the lysosomal fraction of homogenates from many different tissues (Patterson *et al.*, 1965; Mahadevan and Tappel, 1967; McDonald *et al.*, 1968; Tappel, 1969). The pH optimum of the dipeptidyl arylamidase II from the pituitary was pH 5·5; that of liver arylamidase was 6·8, as was that of the arginyl-arylamidase from the pituitary (Ellis and Perry, 1966). The last-named enzyme was much less inhibited by puromycin than was lysyl arylamidase; the antibiotic seemed to act as a competitive inhibitor. McDonald *et al.* (1968) showed that the dipeptidyl arylamidase II was inhibited by Tris buffer; greater inhibition was caused by puromycin aminonucleoside and yet more by puromycin itself, i.e. the larger the cation the greater the inhibition. In contrast, cathepsins B and D are unaffected by puromycin.

It is too early to be decisive as to the significance of the dipeptidyl and tripeptidyl arylamidases, the arginyl arylamidases and the other arylamidases which act on β-naphthylamides of single amino-acids. It may be that they show considerable overlap of specificity. It does seem clear, however, that many are lysosomal enzymes, i.e. they occur in the lysosomal fraction and they show latency. There is some strong suggestion that other enzymes may also have arylamidase (naphthylamidase) activity. Snellman (1969) purified cathepsin B 200 times (over the crude extract) and showed that it was active against leucine 2-naphthylamide. These results were in agreement with a long series of studies which unified data from biochemical and histochemical investigations (Sylvén and Bois, 1963; Sylvén and Bois-Svensson, 1964; Sylvén, 1968). For example, Sylvén and Snellman (1962) showed that cathepsin C, which normally does not hydrolyse leucine amide, could do so under suitable conditions so implying that both types of activity could occur together (also see the biochemical data on proteases, above).

HISTOCHEMICAL DATA

Histochemically it seems that the naphthylamidase reaction, namely the release of naphthylamine from an amino-acid 2-naphthylamide (e.g. leucine β-naphthylamide), is produced by cytoplasmic aminopeptidases and by other predominantly lysosomal naphthylamidases. Thus the localization of the stain will depend on the relative concentrations of these enzymes in the cells and on their relative rates of action on the synthetic substrate used (see above). In fact naphthylamidase activity has been observed to occur mainly in lysosomal particles (Sylvén and Bois-Svensson, 1964; McCabe, 1964; McCabe and Chayen, 1965; Sylvén and Lippi, 1965; Sylvén, 1968). The reaction has been at the same sites as are coloured by the acid phosphatase reaction, both in autophagic vacuoles (Niemi and Sylvén, 1969) and in lysosomes (Chayen *et al.*, 1971a). Moreover, the naphthylamidase reaction, measured quantitatively, showed the characteristics of latency known for lysosomal membranes (Bitensky *et al.*, 1972). The quantitative histochemical study of lysosomal naphthylamidase has proved to be of considerable value in the study of human rheumatoid arthritis (Chayen and Bitensky, 1971;

Bitensky *et al.*, 1972) and for the investigation of inflammation and of the potency of anti-inflammatory drugs (Chayen *et al.*, 1970). There is also considerable evidence that this enzymic activity is significant in many pathological conditions. Fibrocytes have particularly strong naphthylamidase activity. Burstone (1962) reviewed this whole subject and the considerable literature on the use of this reaction for studying malignant growths. Sylvén and Malmgren (1957) related the invasiveness of the growths to the degree of activity of this enzyme and suggested that aminopeptidase activity was involved in the extracellular proteolysis which must occur during malignant invasion. Although the inflammatory response of the host tissue involves high naphthylamidase activity, the invading tissue seems to contribute its own quota of this activity. Willighagen and Planteijdt (1959) suggested that this activity could be used diagnostically. The work of Braun-Falco on aminopeptidase activity of skin in various pathological conditions was well reviewed by Burstone (1962).

Rationale of the histochemical procedure

(*i*) *The reaction*

$$(H_3C)_2CH-CH_2-CH(NH_2)-C(=O)\!\uparrow\!-NH-C_{10}H_7$$

leucyl naphthylamide

The arrow marks the amide bond

$$\xrightarrow{+\,H_2O}\ (H_3C)_2CH-CH_2-CH(NH_2)-C(=O)OH\ +\ H_2N-C_{10}H_7$$

leucine β-naphthylamine

The enzyme liberates naphthylamine and this has to be trapped, with a diazonium salt, to form an insoluble azo dye (see Sylvén, 1970). In studies on rates and distances of diffusion (Nachlas *et al.*, 1957a; McCabe and Chayen, 1965) Fast Blue B has been shown to be more effective in reducing the diffusion of the naphthylamine than either Fast Garnet GBC or Corinth V. In order that the coupling reaction should be fully effective, it is important that the coupler should be fresh; it has been shown spectrophotometrically (Poulter, unpublished data) that Fast Blue B decomposes in the incubation medium and so must be replenished every 15 min if its effective trapping concentration is to be maintained. In practice it is easier to replace the whole medium, fresh, every 15 min; this is the 'pulsing' technique of Poulter (unpublished data).

(*ii*) *pH 'optimum'*. Originally, when it was thought that the naphthylamidase reaction demonstrated the cytoplasmic aminopeptidases, like those of the kidney, it was necessary to emphasize that the pH of the histochemical reaction (pH 6·5) was a compromise between that giving optimal enzymic activity and that giving adequate coupling of the liberated naphthylamine. Moreover diazonium salts, used for coupling, are more inhibitory to aminopeptidases at the more alkaline pH values (see Barka and Anderson, 1963, p. 288; Nachlas *et al.*, 1957a). Sylvén (1968) has always maintained that if the reaction is done at pH 5·5 there is no residual aminopeptidase activity and the activity demonstrated can be solely that of other naphthylamidases and possibly of cathepsin B. It now seems that pH 6·5 may be optimal for lysosomal arylamidases.

(*iii*) *Latency*. Where the enzyme occurs inside lysosome-like particles, the activity can be enhanced by increasing the permeability of the membrane which bounds the particles. This allows more substrate to come into contact with the enzyme in unit time. It can be effected by treating the section with 0·05M acetate buffer, pH 5·0, at 37°C for varying periods (2–15 min, depending on the tissue) prior to the normal incubation procedure. (Other methods were given by McCabe and Chayen, 1965.) Comparison of the 'freely available' activity (i.e. the normal reaction) and of the 'total' activity (i.e. the maximal activity which the acidic pre-incubation can produce) is a valuable indication of the state of the lysosomal membranes. This is especially useful if the activities are measured by means of a microdensitometer. Then the proportion of 'bound' or latent activity (total-free/total) is an index of the impermeability, and thus of the stability, of the lysosomal membranes. (For full details, including the rigorous use of microdensitometry needed for such studies, see Bitensky *et al.*, 1972). Rosenblatt *et al.* (1958) have suggested that leucyl-4-methoxy-β-naphthylamide may be a better substrate in that the methoxy-naphthylamine couples very much more rapidly than does β-naphthylamine. In our experience lysosomal membranes appear to be completely permeable to this substrate so that it becomes impossible to assess their state with this substrate.

(*iv*) *Activators and inhibitors*. Low concentrations of cyanide may activate the naphthylamidase reaction although higher concentrations inhibit. Other inhibitors include zinc (which is often present in commercially available diazonium salts), copper, mercury, lead and nickel (see Burstone, 1962). Sylvén has studied various activators and inhibitors which may discriminate between 'aminopeptidase' and 'catheptic' activities (e.g. Sylvén, 1968).

METHOD (pulsing procedure of Poulter; also see McCabe and Chayen, 1965; Nachlas *et al.*, 1957a)

Fresh cryostat sections should be used.

1. Incubate the section in the reaction medium in a Coplin jar at 37°C for 15 min.
2. Remove the reaction medium and replace with fresh medium for another 15 min. Repeat this process until sufficient reaction is obtained.

   [*Note:* 8 μ sections of rat kidney produce a strong reaction in the first 15 min whereas those of rat liver require 2 hr to produce sufficient reaction.]
3. Rinse well in an 0·85% solution of sodium chloride at room temperature.
4. Transfer to an 0·1M solution of copper sulphate (2·5 g $CuSO_4.5H_2O$ Analar reagent or its equivalent in 100 ml distilled water) at room temperature for 2 min.

   [*Note:* this step is to chelate, and so intensify and further insolubilize, the dye that has been formed. The localization is unaltered, as was shown by McCabe and Chayen, 1965.]
5. Mount directly in Farrants' medium.

Result. Blue lysosome-like particles or purple stain. The stain fades gradually after about a week. It can be partly restored by reimmersion in copper sulphate solution.

Control. There is no adequate control. Sections can be incubated in the absence of substrate; alternatively the full incubation medium can be used on sections in which naphthylamidase activity has been inhibited by treatment with absolute ethyl alcohol at 37°C for 30 min.

Reaction medium

Prepare fresh:

0·1M acetate buffer, pH 6·1	10 ml
0·85% solution of sodium chloride	8 ml
0·02M (32 mg in 25 ml water) of potassium cyanide	1·0 ml
Leucyl-β-naphthylamide	8 mg in 1·0 ml of distilled water
Just before use add: Fast Blue B	10 mg

Adjust the final pH to pH 6·5.

β-GLUCURONIDASE

BIOCHEMICAL DATA

This enzyme (Enzyme Commission No. 3.2.1.31) hydrolyses β-D-glucosiduronic and β-D-galactosiduronic acids; it does not act on glucosides or on α-D-glucosiduronic acids (*Biochemists' Handbook*, p. 227).

COOH — O — OR $\xrightarrow{+ H_2O}$ COOH — O — HOH + ROH

Its main physiological function is generally believed to be the hydrolysis of steroid glucosiduronic acids (where R is the steroid moiety) and of acidic mucopolysaccharides (*Biochemists' Handbook*). Although it does not catalyse the usual synthesis of glucosiduronic acids, it seems to be capable of transferase activity: i.e.

$R.O.C_6H_9O_6$	+	R'OH	⟶	$R'.O.C_6H_9O_6$	+	R.OH
glucuronic acid (donor substrate)		acceptor (which can be a steroid or polysaccharide like hyaluronic acid)		new glucosiduronic acid (e.g. steroid glucuronide)		aglycone of donor substrate

Fishman (1961; also see Fishman and Baker, 1956) has shown that at physiological pH and optimal concentrations of substrate and acceptor β-glucuronidase action is of this transfer type.

This enzyme seems to occur in most tissues of mammals and has been reported in high concentrations in liver, kidney, spleen, lung, adrenal, thyroid and uterus. In the uterus, vagina, mammary gland, preputial gland, adrenal, liver and kidney, it has been shown to be markedly affected by gonadal hormones (refs. in Fishman, 1963). Within cells its location seems to be mainly in the mitochondrial (or lysosomal) and microsomal fractions (Dixon and Webb, 1964; Roodyn, 1965; Fishman *et al.*, 1964).

The hydrolytic activity of β-glucuronidase varies from pH 3·5–5·0. The enzyme is said to dissociate readily into inactive components; deoxyribonucleic acid, protamine and diamines reverse this dissociation and hence frequently one or other is added as an activator (Bernfeld *et al.*, 1954;

Fialkow and Fishman, 1961). Natural activators which are endogenous in some tissues have also been reported (Fialkow and Fishman, 1961). A remarkable phenomenon has been the greatly enhanced activity of kidney β-glucuronidase after parenteral administration of testosterone-like steroids to mice (Fishman and Lipkind, 1958; Fishman, 1961).

The specific inhibitor of β-glucuronidase is D-gluco-saccharo-1:4-lactone present in saccharate solution; tissue endogenous inhibitors have also been reported. It is inhibited by copper, especially in the presence of ascorbate but not by fluoride, cyanide or thiourea (*Biochemists' Handbook*).

A great number of careful studies indicate clearly that β-glucuronidase activity is very much greater in neoplasms than in the adjacent normal tissue (e.g. Fishman and Anlyan, 1947, 1950). Fishman and Bigelow (1950) showed that this enzyme was very much more active in gastro-intestinal tumours than in homologous human tissue. The work of Kasdon *et al.* (1950, 1951, 1953) drew attention to the elevated levels of this enzyme in human cervical cancer and led to a number of attempts to use this as a diagnostic criterion for distinguishing early malignancy (e.g. Odell and Burt, 1949; Odell *et al.*, 1949); unfortunately the results are not decisive (see Fishman, 1951; Lawson and Watkins, 1965).

HISTOCHEMICAL DATA

This enzyme forms a critical test of histochemistry in that it is readily inhibited by fixatives (see Appendix, p. 239; Barka and Anderson, 1963, p. 281; Nachlas *et al.*, 1956) and by some diazonium compounds (Fishman and Goldman, 1965); it seems to be held in a labile form in tissues (as indicated by its reputed distribution in the various cell fractions; also see Barka and Anderson, 1963, p. 281); and its pH optimum is in the acidic range (3·5–5·5 according to Fishman and Goldman, 1965) at which diazonium compounds do not couple efficiently (see Barka, 1960, for the coupling of hexazonium pararosaniline and Burstone, 1962, pp. 95–108 for full discussion of diazonium coupling reactions).

The considerable controversies that have raged over the various methods for demonstrating this enzyme reflect some of these difficulties. The method of Fishman and Baker (1956) was much used and has been recommended by Barka and Anderson (1963); it has become displaced by the newer diazonium coupling methods partly because the complex substrates are now available commercially.

The histochemistry involves the following steps:

1. The labile enzyme must be retained at its natural site during the preparation of the sections and it should not be inhibited. This becomes especially serious because it is not beyond belief that any given inhibition may act solely on a microsomal or solely on a lysosomal enzyme (if there are truly two β-glucuronidases in cells).

2. The readily soluble enzyme (Barka and Anderson, 1963, p. 281) must be retained at its natural site during the histochemical incubation procedure.

3. The enzyme then is allowed to act on a synthetic glucuronide and to split off from it a naphthol derivative.

Hayashi *et al.* (1964) used a naphthol AS-BI glucuronide in the presence of hexazonium pararosaniline (cf. method for acid phosphatase). In this simultaneous coupling reaction the coupler (hexazonium pararosaniline) is expected to trap the liberated naphthol AS-BI as it is freed from the glucuronide. A compromise of pH 5·2 was used because, even at this pH, the coupling is not very efficient (Barka, 1960) but at higher pH values the enzyme is very much less active.

These considerations led Fishman and Goldman (1965) to use a post-coupling method. The naphthol AS-BI is liberated by the enzyme at a pH which is optimal for the enzyme, i.e. at pH 4·5, in the absence of inhibitory diazonium compounds. The section is then transferred to a solution containing the most efficient coupler (irrespective of whether it is inhibitory or not because the enzymatic reaction has now been completed) at the pH which is optimal for the coupling reaction. The whole process, of course, depends on the naphthol AS-BI being deposited, and remaining, at the site of its liberation.

Another advantage of this type of procedure is that the activity of the enzyme can be studied at various pH values without having the variable coupling potency of the coupler, at these pH values, to confuse interpretation of the results.

The histochemistry of this enzyme has been studied in some detail by the present authors who found that the enzyme is readily liberated into the general cytoplasm which becomes coagulated during fixation; the enzyme-rich granules, in such a section, appeared to be denser aggregates of the cytoplasmic-diffuse enzyme coagulum, with the enzyme also distributed generally throughout the cytoplasm. When care was taken to ensure that the cytoplasm did not coagulate, the enzyme was found to be located solely in discrete granules which could be activated by as short as one minute exposure to acidic conditions (in the absence of stabilizer).

Because these studies are not yet published, both the simultaneous and the post-coupling methods will be given as well as the newer modification.

SIMULTANEOUS COUPLING METHOD (Hayashi *et al*, 1964)

METHOD

1. Fix thin slices of tissue (5×5×3 mm) in formol calcium (4% w/v formaldehyde and 1% anhydrous calcium chloride) at 4°C for 24 hr.

2. Do not wash; only blot on filter paper. Transfer to Holt's hypertonic gum-sucrose medium (0·88M sucrose containing 1% gum acacia) for at least 24 hr at 4°C.

[*Note:* Holt, 1959, showed that this treatment removed much of the inhibitory effect of the formalin fixation; however it does not appear to allow the sub-cellular membranes to revert to their normal permeability.]

3. Cut frozen sections of this fixed tissue. Preferably use free-floating sections.
4. Incubate at 37°C in the reaction medium for 20–30 min (for rat kidney and liver).
5. Rinse well with distilled water.
6. Mount the free-floating sections on to slides.
7. Dehydrate in alcohol, clear in xylol and mount in Technicon mounting medium (or any suitable mountant).

Result. Red stain, usually or predominantly in small granules, shows enzyme activity.

Solutions required for this method

Substrate stock solution

Dissolve 28 mg of naphthol AS-BI glucuronide in 1·2 ml of 0·05M sodium bicarbonate (0·42 g sodium bicarbonate in 100 ml of water). Add 0·2M acetate buffer at pH 5·0 to give a final volume of 100 ml. This stock solution can be kept at room temperature for a few weeks.

Pararosaniline solution

Dissolve, warming gently, 1 g of pararosaniline hydrochloride in acidified distilled water (20 ml distilled water + 5 ml of concentrated hydrochloric acid). Cool; filter and store at room temperature.

Sodium nitrite solution

4% sodium nitrite in distilled water. Store in a refrigerator. This should not be used if kept for longer than one week.

Reaction medium

First add 0·3 ml of pararosaniline solution to 0·3 ml of the sodium nitrite solution in a 20 ml beaker. Then, after 1 min, add 10 ml of the substrate stock solution.

With caustic soda solution, adjust the pH to 5·2. After this, make the volume of the final solution up to 20 ml with distilled water. Filter through a Whatman No. 1 filter paper.

POST-COUPLING METHOD (Fishman and Goldman, 1965)

METHOD

Fishman and Goldman recommended either frozen sections of fixed tissue (as for the simultaneous coupling method) or cryostat sections of fixed or of unfixed tissue.

1. Add the reaction medium to the section at 37°C at optimal pH. 30 min incubation should be sufficient for most tissues; 60 min will be required for tissues containing low β-glucuronidase activity. In general the pH will be 4·5 (as found for β-glucuronidase purified from rat liver).
2. Wash in cold distilled water.
3. Immerse in a cold saturated solution of the diazonium coupling agent in phosphate buffer at pH 7·4. Leave (e.g. in a refrigerator) for 2–5 min with gentle agitation.
4. Wash in distilled water.
5. Mount in 1% aqueous polyvinylpyrrolidone.

Result. As with the simultaneous coupling method. The colour will depend on the coupler used.

Solutions required for this method

Substrate stock solution
Dissolve 11 mg of naphthol AS-BI-β-D-glucosiduronic acid in 1·0 ml of 0·05M sodium bicarbonate; make the volume of this solution up to 100 ml with 0·1M acetate buffer of the required pH (e.g. pH 4·5). This solution is stable for months in a refrigerator; it is 2×10^{-4}M with respect to the substrate. Optimum substrate concentration is said to be 1×10^{-4}M.

Reaction medium
Dilute the stock either to 1×10^{-4}M (i.e. 1 : 1) or to 4×10^{-5}M (for economy) with 0·1M acetate buffer, pH 4·5.

Diazonium coupler
A saturated solution of any suitable diazonium salt in 0·01M phosphate buffer, pH 7·4, can be used. They list, as 'suitable', Fast Garnet GBC (Gurr); Fast Dark Blue R (Rohner Ltd.), Fast Blue RR (Dajac); Fast Blue 2B (National Aniline); Fast red violet 2B (Sigma).

MODIFIED POST-COUPLING METHOD

The procedure which we prefer is as follows:

METHOD

1. Use unfixed fresh cryostat sections. Leave at room temperature only so long as is required to ensure that the surface moisture has gone.
2. Immerse in the reaction medium at 37°C in a Coplin jar in a water bath. Leave for 30 min to 3 hr depending on the tissue. Rat liver required about 2 hr incubation unless it had been activated (see below).

3. Wash in 0·1M acetate buffer, pH 4·5, which contains 0·5% of calcium chloride ($CaCl_2 . 6H_2O$).

[*Note:* calcium was found to be essential for retaining the integrity of the sections and of the reaction.]

4. Immerse for 5 min in the cold coupling solution in a refrigerator.
5. Rinse in distilled water at room temperature.
6. Mount in glycerine jelly which has been warmed to 40°C to render it sufficiently fluid.

[*Note:* in our studies the mountant seemed to be fairly critical. For this reaction, Farrants' medium was less satisfactory. It was possible to see changes in the distribution of the dye with prolonged immersion in some mountants.]

Result. Dark blue reaction, generally in discrete granules, indicated β-glucuronidase activity; they were absent in controls (see below).

Solutions required for this method

Stock substrate solution

11·4 mg naphthol AS-BI glucuronide are dissolved in 1 ml of an 0·05M solution of sodium bicarbonate (0·42 g sodium bicarbonate in 100 ml of distilled water). To this is added 49 ml of an 0·2M acetate buffer at pH 4·5. This stock solution was stored indefinitely at +4°C.

PVA solution

Dissolve 0·5 g of calcium chloride ($CaCl_2 . 6H_2O$) in 100 ml of an 0·1M acetate buffer, pH 4·5. Add 10 g of polyvinyl alcohol (PVA: see Appendix for details). It is not readily soluble and the solution has to be heated—but not boiled—with frequent mixing for about 5 min to dissolve the PVA. It is cooled before it is used.

Reaction medium

PVA solution	20 ml
Stock substrate solution	2 ml

Coupling solution

Dissolve 8 mg Fast Dark Blue R (Gurr) in 20 ml of an 0·01M phosphate buffer (disodium hydrogen phosphate: potassium dihydrogen phosphate), pH 7·4 at +4°C.

This solution, at +4°C, is filtered through a Whatman No. 1 filter paper and should be used immediately.

Glycerine jelly

Dissolve 15 g white gelatine in 100 ml of distilled water (with gentle heating if necessary). Add 100 g glycerine and warm for 5 min in a water bath at 37°C. Filter the warm solution through glass wool.

This jelly should be stored frozen in small quantities.

Control. 0·05 g potassium hydrogen saccharate (B.D.H.) is added to 22 ml of the reaction medium and the sections are incubated directly in this. Complete inhibition of β-glucuronidase activity has been observed with this procedure.

Activation. Immerse sections in a 0·1M acetate buffer, pH 4·5, which contains 0·5% calcium chloride ($CaCl_2 . 6H_2O$). Leave for 2 min and then transfer to the reaction medium (step 2, above). Continue as for the normal reaction.

PHOSPHORYLASES

Phosphorylase activity is the culmination of a number of enzymic reactions. The biochemistry of each enzyme involved will be discussed separately before considering the histochemistry of 'phosphorylase activity'. It should also be noted that many enzymes show transglycosilase activity, i.e. they will transfer a sugar moiety. The true phosphorylases transfer glycosyl groups to and from phosphate itself; other transglycosylases (for example, that enzyme listed by the Enzyme Commission No. 2.4.1.7) can act as phosphorylases but in fact can use a number of acceptors other than ortho-phosphate (e.g. some enzymes transfer parts of glycogen to parts of amylopectin to form a branched structure; others use uridine diphosphate as the acceptor; others use disaccharides, etc., as discussed by Dixon and Webb, 1964, p. 188).

PHOSPHORYLASE a

The best known phosphorylase enzyme is glycogen phosphorylase, which is also called α-glucan phosphorylase (Enzyme Commission No. 2.4.1.1); it catalyses the reversible reaction:

$$\text{Glycogen} + n\text{H}_3\text{PO}_4 \rightleftharpoons n(\text{glucose-1-phosphate})$$

The equilibrium is pH-dependent and favours the synthesis of glycogen, i.e.

$$n(\text{glucose-1-phosphate}) + (\text{glycogen})_m \rightarrow (\text{glycogen})_{m+n} + n\text{H}_3\text{PO}_4$$

where $(\text{glycogen})_m$ represents the primer of chain length m which becomes extended, by the phosphorylase activity, to a chain length of $m+n$. Thus it will be noted that the effect of this enzyme, in this reaction, is to transfer glucose residues from glucose-1-phosphate to the primer carbohydrate chain.

For maximum activity adenosine 5-phosphate (AMP) is required as the main activator (in its absence the activity drops to 65% of maximal) but traces of cysteine or of EDTA are also necessary. The pH optimum is 6·1–6·9. Glucose is a competitive inhibitor; phlorhizin and *p*-chloromercuribenzoate are non-competitive inhibitors (see *Biochemists' Handbook*, p. 479).

Phosphorylase activity is found predominantly in the supernatant fraction (Dixon and Webb, 1964).

It has been reported that it occurs in the microsomal fraction in well-fed rats but is released into the supernatant fraction when the rats are starved (see Tata, 1964, quoted by Roodyn, 1965).

Phosphorylase *a* has a molecular weight of 495,000 and contains four molecules of pyridoxal phosphate (which do not participate directly in the enzymic activity) per molecule of the enzyme. It has four active sites per molecule and can bind one molecule of AMP to a serine residue at each active site (hence the effect of AMP as an activator).

PHOSPHORYLASE b

Some, or much, of the potential phosphorylase activity of tissues can be present in an enzymatically inactive form, known as phosphorylase *b*. This has a molecular weight of 242,000; contains two molecules of pyridoxal phosphate per molecule of the enzyme and can bind two molecules of AMP per molecule of phosphorylase *b*. In the presence of magnesium ions it forms a dimer (i.e. two molecules of phosphorylase *b* linked together). However its conversion to phosphorylase *a* (and hence to the enzymatically active form) requires enzymatic phosphorylation of the phosphorylase *b*; this is effected by a specific enzyme 'phosphorylase kinase' (Enzyme Commission No. 2.7.1.38) which adds two phosphate groups to each molecule of phosphorylase *b*.

Phosphorylases appear to play a major role in controlling the metabolism of glycogen in animal cells generally; they are found in highest concentration in voluntary and heart muscle and in the liver (*Biochemists' Handbook*).

INTER-CONVERSION OF PHOSPHORYLASES a AND b

$$\underset{\text{inactive}}{\text{Phosphorylase } b} + Mg^{2+} + 2ATP \xrightarrow[\text{(2.7.1.38)}]{\text{kinase}} \underset{\text{active}}{\text{phosphorylase } a}$$

The phosphorylase kinase which causes this reaction is stimulated by epinephrine and by glucagon (*Biochemists' Handbook*). In the presence of manganese or magnesium ions (the former seems to have been used in studies on muscle and the latter on liver) the kinase transfers two phosphate groups (from ATP) to each molecule of the inactive phosphorylase *b* which, in the presence of Mg^{2+} or Mn^{2+}, forms dimeric phosphorylated molecules of active phosphorylase *a*.

$$\text{Phosphorylase } a \xrightarrow[\substack{\text{phosphatase} \\ \text{(3.1.3.17)}}]{\text{phosphorylase}} 2(\text{phosphorylase } b) + 2H_3PO_4$$

The phosphorylase phosphatase is found in the soluble fraction of homogenates (*Biochemists' Handbook*).

HISTOCHEMICAL DATA

The particular interest in this enzyme has been its remarkably decreased activity in tumours, as shown in a long series of detailed studies by Godleswki (e.g. 1962, 1963). There has also been some discussion of whether activity of the 'branching enzyme' (mentioned in the biochemical data) can be seen in histochemical phosphorylase reactions by its products giving a violet-red or brown-purple colour with the iodine used in the reaction (see Burstone, 1962, p. 368, discussing work by Takeuchi).

In the histochemical method the section is given glucose-1-phosphate and a little glycogen; the tissue phosphorylase is encouraged to increase the size of the glycogen by accretion of glucose units from the glucose-1-phosphate (as in the 'equations' above). This reaction depends on the presence of the active enzyme, phosphorylase *a* and will be reduced if that enzyme is simultaneously being dephosphorylated, by the phosphorylase phosphatase, to the inactive form. Consequently fluoride is added to the incubation medium to inhibit the action of this phosphatase. It is hoped that the extended molecules of glycogen will become bound to the section and they are then demonstrated by their response to iodine. It is claimed that, with iodine, linear amylose stains blue but branched polysaccharide like glycogen stains red-brown. Consequently, if phosphorylase activity alone has occurred, the end-product should stain blue whereas if the phosphorylase has worked in conjunction with the branching enzyme, then a purple (blue + red-brown) colour should result, It has been found by Poulter and Bitensky (unpublished data) that the binding of the glycogen to the section is greatly enhanced by leaving the section at room temperature for 10 min before it is incubated. Insulin may be added to the incubation medium to enhance phosphorylase activity (Barka and Anderson, 1963, p. 291).

If the total potential phosphorylase activity is required, it becomes necessary to convert the inactive phosphorylase *b* into the active form (phosphorylase *a*). This conversion must be effected enzymatically (see above); it involves the addition of phosphate (from ATP) to the phosphorylase *b*, by phosphorylase kinase, and the dimerization of phosphorylase *b* molecules by the action of magnesium ions. Hence to demonstrate total potential activity, ATP and Mg^{2+} are added to aid the conversion of phosphorylase *b* to phosphorylase *a*, and fluoride is added to stop the degradation of the latter.

METHOD* (after Takeuchi and Kuriaki, 1953; Godlewski, 1960, 1964)

Fresh cryostat sections, preferably 20 μ thick, should be used. They should be kept at +4°C for 1–3 hr, or left to dry at room temperature for 10 min, before use.

1. Incubate in the selected test medium. For some tissues the incubation may be 20 min at 37°C; others may require 60 min at 30°C.

   [*Note:* the medium selected, I–V below, will depend on whether the activity that is to be studied is that which is already manifest in the specimen, or the total activity of which the specimen is capable.]

2. Wash briefly in distilled water.
3. Immerse in Gram's iodine, diluted 1:2 with distilled water, for 30 sec or up to 5 min.
4. Mount in glycerine–iodine mixture.

* We are grateful to Mr. A. Sullivan for his help with this method.

Solutions required for this method

Medium:	I	II	III	IV	V
Enzymes demonstrated:	P_a, P_b and BE	P_a and P_b	P_a and BE	P_a only	Control
G-1-P (mg)	50	50	50	50	—
5′AMP (mg)	10	10	10	10	10
Glycogen (mg)	5	5	5	5	5
EDTA (mg)	10	10	—	—	10
G-6-P (mg)	—	—	—	—	50
NaF (mg)	—	—	52·5	52·5	—
0·5M Acetate buffer, pH 6·0 (ml)	10	10	10	10	10
H_2O (ml)	15	2·5*	15	2·5*	15
40% Ethanol (ml)	—	12·5*	—	12·5*	—

* Or replace the water and ethanol by 1 mg $HgCl_2$ in 15 ml of water.

P_a = phosphorylase *a*,
P_b = phosphorylase *b*,
BE = branching enzyme,
G-1-P = glucose-1-phosphate,
G-6-P = glucose-6-phosphate.

Reaction medium for total potential phosphorylase activity

Use reaction medium III but add 10 mg ATP and 10 mg of magnesium sulphate to activate phosphorylase kinase (PK). The presence of sodium fluoride should inhibit the breakdown of phosphorylase *a* to phosphorylase *b* by phosphorylase phosphatase (PPase).

Glycerine–iodine mixture

One volume of glycerine to 1 volume of Gram's iodine.

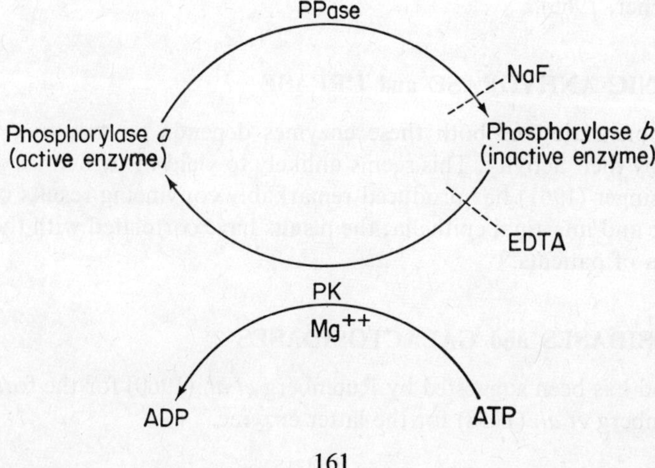

Results. Mediums I and III demonstrate phosphorylase activity and branching enzyme, the former producing a blue colour and the latter an orange-red colour.

Mediums II and IV show phosphorylase activity only, without branching enzyme.

The colours produced fade in 12–24 hr.

Other controls. For controls for branching enzyme see Burstone (1962) and Barka and Anderson (1963).

Other controls consist of pretreating sections with diastase (or saliva) to remove endogenous glycogen; any glycogen found at the end of the reaction is then solely that which has been formed, and bound to the section, during the course of the reaction. Diastase (see p. 79) can also be used at the end of the reaction to prove that the colour is indeed due to glycogen.

Certain other enzymes deserve mention, but because the methods are not widely used, full details will not be given. Anyone concerned with the study of these enzymes is asked to refer to the papers cited.

SULPHATASES

Biochemically these enzymes (e.g. arylsulphatase, Enzyme Commission No. 3.1.6.1) appear to be located either in the 'mitochondrial' (lysosomal) or in the microsomal fraction (see Roodyn, 1965; *Biochemists' Handbook*, pp. 270–71). Probably the most promising histochemical method for these enzymes is that developed by Ostrowski (Kawiak *et al.*, 1964). Other methods include those of Rutenberg *et al.* (1952; also see Rutenberg and Seligman, 1956); the use of naphthol AS-BI-sulphate (Woohsman and Hartrodt, 1964) used as for the equivalent glucuronidase reaction (see p. 153); and the use of lead to precipitate the sulphate ions liberated from nitrocatechol sulphate (Goldfischer, 1965).

CARBONIC ANHYDRASE and UREASE

The demonstration of both these enzymes depends on trapping the CO_2 evolved by their activity. This seems unlikely to yield histochemical methods but Freisinger (1961) has produced remarkably convincing results on urease in gastric and intestinal epithelia; the results have correlated with the clinical condition of patients.

GLUCOSIDASES and GALACTOSIDASES

A method has been suggested by Rutenberg *et al.* (1960) for the former, and by Rutenberg *et al.* (1958) for the latter enzyme.

OXIDASES

Oxidases are enzymes which catalyse the oxidation of a substrate by removing hydrogen from it and passing it to an acceptor which can be oxygen. This is in contrast to the usual dehydrogenases which also remove hydrogen from their specific substrate but donate it to another acceptor which cannot be oxygen. Thus oxidases and dehydrogenases are both classified as 'oxido-reductases' in that they catalyse the same general reaction:

$$SH_2 + A \longrightarrow S + AH_2$$

(where S is the substrate; SH_2 is the reduced substrate; A is the acceptor and AH_2 is the reduced acceptor).

For oxidases, A is oxygen, or an acceptor which has an electrode potential equivalent to that of oxygen. For dehydrogenases, A is a hydrogen-acceptor of an electrode-potential nearer to that of free hydrogen than the electrode-potential of oxygen. This concept is so fundamental to the histochemistry of oxidative reactions that some discussion of it must be given in this book (see under 'dehydrogenases'). However, in this section, we will consider those oxido-reductases which use either oxygen, or dyes which can replace oxygen, as the hydrogen acceptor.

At present considerable advances are being made in the histochemistry of oxidases and peroxidases based on the use of diamino-benzidine as developed initially by Graham and Karnovsky (1966). Although these developments have been made mainly for the ultrastructural localization of such enzymes they may also be of great value for light microscopical histochemistry (e.g. Novikoff and Goldfischer, 1969; Goldfischer and Essner, 1969). Reference should be made to Seligman *et al.* (1968, 1970a and b). Shnitka and Seligman (1971) have reviewed these methods and newer electron-histochemical techniques for dehydrogenases (and other enzymes).

CYTOCHROME OXIDASE

BIOCHEMICAL DATA

Cytochromes were so named because they were 'cell pigments'. Originally three were described and named cytochromes *a*, *b* and *c*. As each new cytochrome has been discovered, so it has been named according to which of the original three it most closely resembles. Thus today, each cytochrome genus

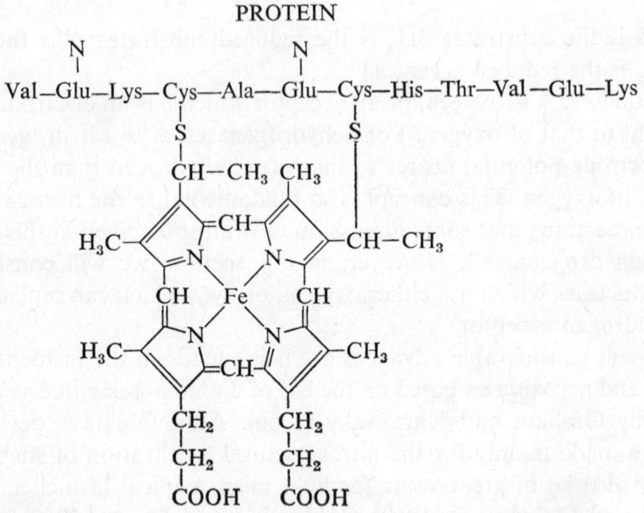

PROTEIN

N N

Val—Glu—Lys—Cys—Ala—Glu—Cys—His—Thr—Val—Glu—Lys

S S

CH—CH_3 CH_3

H_3C =CH— CH—CH_3

N N

CH Fe CH

N N

H_3C =CH— CH_3

CH_2 CH_2

CH_2 CH_2

COOH COOH

Fig. 20. Linkage of porphyrin to protein in cytochrome *c*.

has many species, which are distinguished by a subscript, i.e. for cytochrome *a* (genus) there are cytochromes a_1, a_2, a_3 and a_4 (species). Some species of cytochrome have been found in one or other organism; in mammals the cytochromes are a_3, *a*, *c*, c_1, *b* and b_5 (West and Todd, 1961, p. 853). The standard electrode potential, E_0', of the cytochrome *c* system is +0·26 volt at pH 7 (West and Todd, 1961, p. 853). There is some doubt whether cytochromes *a* and a_3 can be clearly distinguished (Dixon and Webb, 1964; Lehninger 1965, p. 77); cytochrome oxidase is cytochrome *a*, with or without cytochrome a_3 (Dixon and Webb, 1964, p. 391). Active cytochrome oxidase is a complex between cytochromes *c* and *a* (Dixon and Webb, 1964, p. 389):

reduced cytochrome *c* does not react with oxygen at neutral pH values unless cytochrome oxidase is present. Similarly cytochrome oxidase cannot oxidize *p*-phenylenediamine, or cysteine or ascorbate in the absence of cytochrome *c*.

The cytochromes consist of a particular porphyrin linked to protein. Iron is present in the porphyrin ring, but in some, particularly in cytochrome oxidase, copper appears to be an integral part of the enzyme (see West and Todd, 1961, p. 854). The linkage between the porphyrin and the protein has been determined in the case of cytochrome *c* (Fig. 20).

Cytochrome oxidase (also called cytochrome *c* oxidase) seems to form a structural component of mitochondria (see Lehninger, 1965) and consequently its presence or absence is often used to test to what extent other cellular fractions have been contaminated by mitochondria (e.g. de Duve, 1957). It can be 'solubilized' from the mitochondrial fraction by treatment with deoxycholate and trypsin (*Biochemists' Handbook*, p. 382) when it is found to contain surprisingly large amounts of lipid. It is extremely sensitive to cyanide: 10^{-8}M cyanide gives 50% inhibition (in contrast to 80% inhibition given by 10^{-3}M azide) while 10^{-4}M cyanide is usually sufficient to give total inhibition (see Dixon and Webb, 1964; *Biochemists' Handbook*, p. 625). It is also very sensitive to hydrogen sulphide and to carbon monoxide.

HISTOCHEMICAL DATA

This subject has been so bedevilled by odd names and initials that some discussion of these must be given to clarify the nomenclature.

When a naphthol and an aromatic diamine are mixed in the presence of oxygen, a coloured indophenol is produced; this reaction is immediate if a strong oxidizer like hypochlorite is added (Gomori, 1952, pp. 153 *et seq.*):

H_3C–N–CH_3 … NH_2 (aromatic diamine) + OH (naphthol) $\xrightarrow{+ O_2}$ H_3C–N–CH_3 … N= … O (coloured indophenol)

Because this involves a *na*phthol with a *di*amine it was called the *nadi* reaction. When it was discovered that most tissues, if fresh, could take the place of the hypochlorite, that is to say they contained an oxidizing factor (or enzyme) which could produce this indophenol, the 'enzyme' was called indophenol oxidase. It was then discovered that tissues could form the indophenol in two different ways, one of which did not appear to be enzymatic. The reaction which was much investigated in *m*yeloid leucocytes (hence the M-nadi reaction) was insensitive to formalin (even after 6 years according to Gomori, 1952), to alcohol and to acetone; its maximal activity occurred at pH 12·5–13 and it occurred in the absence of oxygen. In contrast, the reaction found in fresh tissues ('tissue' is 'Gewebe' in German: hence this is the G-nadi or Gewebe-nadi reaction) was inhibited by acetone, alkali, acid, formalin, heat (55°C) and it did not occur in the absence of oxygen. Finally it was shown that cytochrome oxidase could be responsible for the G-nadi reaction. It must be appreciated, however, that the ability to produce an indophenol through oxidation of a naphthol–aromatic diamine mixture is not exclusively the prerogative of cytochrome oxidase: this is shown by the fact that it can occur non-enzymatically, as in the M-nadi reaction. Moreover any oxidizing factor, enzymic or otherwise, which can operate at this electrode potential or replace hypochlorite in the original studies, will produce indophenol. In the histochemical method (which owes much to work by Burstone, e.g. 1959, 1960) the equivalent of oxygen is provided by oxidized cytochrome *c*, so giving some degree of specificity to the reaction; the function of the cytochrome oxidase (or more correctly, of the cytochrome *a* part of the cytochrome *c*/cytochrome *a* complex) is to re-oxidize the cytochrome *c* which becomes reduced in forming the indophenol.

N-phenyl-*p*-phenylene diamine + 1-hydroxy-2-naphthoic acid + oxidized cytochrome *c*

the aromatic diamine — the naphthol

cytochrome oxidase + O_2

reduced cytochrome *c* +

indophenol
coloured: note the quinone form

Cytochrome oxidase is of considerable significance in histochemistry. It has been much studied in malignant and benign growths (for a good review see Burstone, 1962). Many workers have found that the cytochrome oxidase activity of many cancers is markedly lower than that of the cells from which the growth originated (also see Butcher *et al.*, 1965).

METHOD (Butcher *et al*, 1964)

Use fresh cryostat sections.

1. Incubate in the reaction medium in a Coplin jar at room temperature [to reduce the rate of spontaneous formation of indophenol]. Rat liver requires up to 60 min incubation; cardiac muscle or kidney are well stained after 20–30 min.
2. Transfer to Lugol's iodine (see p. 79) for 2 min. [This intensifies the colour and stabilizes it against the rapid fading which occurs if iodine is not used.]
3. Treat the section with a 5% aqueous solution of sodium thiosulphate for 4 min to reduce the non-specific iodine background colouration.
4. Place the section in a 1% aqueous solution of cobalt acetate for 60 min at room temperature. [This chelates the dye.]
5. Wash in distilled water.
6. Mount in Farrants' medium.

Result. Almost black reaction which should be restricted to mitochondria.

Control. Incubate as for the normal test but add 10^{-3}M potassium cyanide to the reaction medium (3·25 mg/50 ml).

Solutions required for this method

Reaction medium
Dissolve 10 mg of N-phenyl-*p*-phenylenediamine and 10 mg of 1-hydroxy-2-naphthoic acid in 0·5 ml of absolute ethyl alcohol. Then add 35 ml of distilled water and to this add 15 ml 0·2M Tris-buffer at pH 7·4. Shake. Filter into a Coplin jar.

Tris-buffer, 0·2M, *pH* 7·4
Dissolve 2·42 g Tris in 17 ml 1 N HCl; make up to 100 ml with distilled water.

DOPA-OXIDASE (PHENOLASE)

BIOCHEMICAL DATA

One enzyme, which catalyses the oxidation of both monophenols (e.g. tyrosine) and of ortho-diphenols (such as adrenaline and catechol), is known by any of the following names: phenol oxidase, polyphenol oxidase, phenolase, DOPA oxidase, potato oxidase, catechol oxidase and tyrosinase. The enzyme (Enzyme Commission No. 1.10.3.1) has been purified and studied in detail: it has a molecular weight of 100,000 and contains four atoms of copper in each molecule, the copper being at the active centre of the enzyme and essential for the oxidative catalysis. It is therefore inhibited by substances which complex with copper, including cyanide, diethyldithiocarbamate, glutathione, cysteine, BAL (British Anti-Lewisite) and potassium ethyl xanthate. 4-Nitrocatechol and 4-nitrophenol are competitive inhibitors. Although it is one enzyme, it behaves rather differently as a monophenolase than as a diphenolase. With monophenols there is a lag, or induction, period which can be eliminated or shortened if a trace of catechol or of ascorbic acid is added to the reaction medium (see *Biochemists' Handbook*, pp. 375–76).

This enzyme is very widely distributed throughout the plant and animal kingdoms. Its effect is prominent in that it is involved in the formation of the various melanin pigments. These are complex, high molecular weight polymers, generally combined with protein; all are formed from 3,4-dihydroxyphenylalanine (DOPA) as a result of a series of somewhat complicated reactions (e.g. see West and Todd, 1961, p. 1101). The oxidations in which phenolase is known to be especially significant are of two types:

(*i*) In plants particularly, wounding or cutting often initiates the following reaction (Baldwin, 1959, p. 150):

$$\text{catechol} + \tfrac{1}{2}\,O_2 \xrightarrow{\text{phenolase}} o\text{-quinone} + H_2O \longrightarrow \text{(1,2,4-trihydroxybenzene)}$$

catechol (OH, OH)   *o*-quinone (O, O)   product (OH, OH, OH)

The ortho-quinone reacts with water, apparently spontaneously, and the reaction product is then capable of reacting spontaneously with another molecule of the ortho-quinone to yield hydroxyquinone:

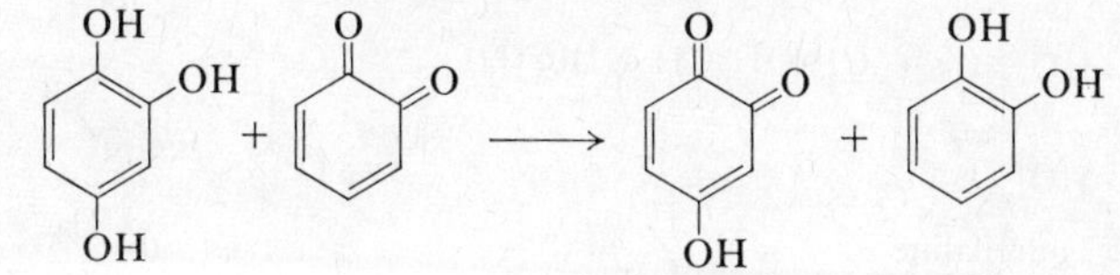

hydroxyquinone catechol

The hydroxyquinone so formed undergoes polymerization to produce complex dark melanin products of unknown constitution.

(*ii*) The formation of melanin from tyrosine begins with the oxidation of tyrosine to form a DOPA quinone; this depends on the monophenol oxidase activity of the enzyme (Baldwin, 1959; West and Todd, 1961, p. 1101):

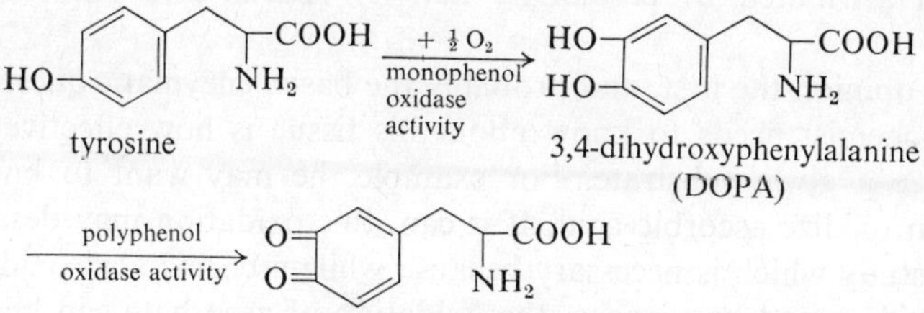

tyrosine $\xrightarrow[\text{monophenol oxidase activity}]{+\frac{1}{2}O_2}$ 3,4-dihydroxyphenylalanine (DOPA)

$\xrightarrow{\text{polyphenol oxidase activity}}$ DOPA-quinone

The DOPA-quinone then undergoes spontaneous ring closure and reaction with water to go through intermediate stages to form first a red pigment and then the indole-5,6-quinone which polymerizes to form melanins (see Baldwin, 1959, p. 151; West and Todd, 1961, p. 1101):

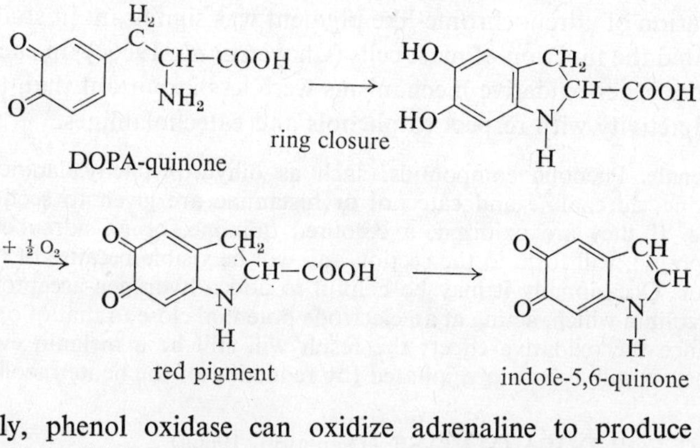

DOPA-quinone $\xrightarrow{\text{ring closure}}$

$\xrightarrow{+\frac{1}{2}O_2}$ red pigment $\longrightarrow$ indole-5,6-quinone

Similarly, phenol oxidase can oxidize adrenaline to produce the red adrenochrome which polymerizes to yield adrenaline-melanin.

Melanins are formed in pigment-forming cells (melanoblasts). The pigments occur in large quantities in tumours of these cells (melanosarcomata); melanin precursors are sometimes found in the urine of patients who have such cancers.

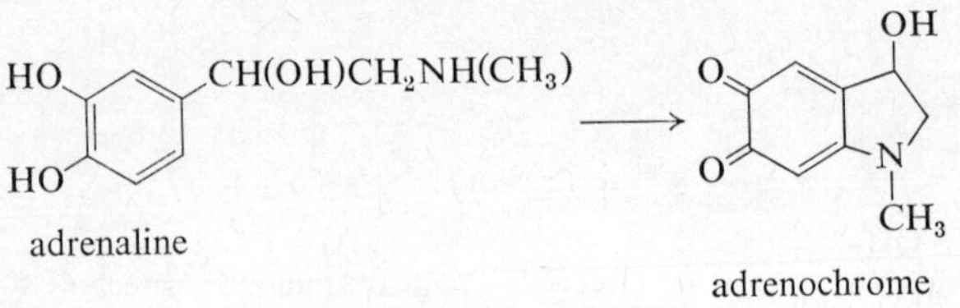

HO, HO — $CH(OH)CH_2NH(CH_3)$ ⟶ O, O, OH, N, CH_3

adrenaline — adrenochrome

HISTOCHEMICAL DATA

The histochemical demonstration of this enzyme may be of diagnostic significance in non-pigmented melanomata (see Burstone, 1962). A 'non-specific DOPA-oxidase' has been much reported in various blood cells and has been attributed to peroxidase activity (Barka and Anderson, 1963, p. 328).

In our opinion the first, and probably the basic, enzymatic question which the histochemist needs to know about his tissue is how effectively does it metabolize a given substrate. For example, he may want to know if the tissue can oxidize ascorbic acid. If it can, this oxidation may deserve more detailed study which is necessary because while ascorbic acid oxidase is not of very widespread occurrence, the oxidation of ascorbate can be mediated by many oxidative enzymes, including phenol oxidase (see Chayen, 1953).

So too with phenols: it is of interest, in the first place, to see if any phenol oxidation takes place; if it does, then the phenomenon may be worthy of closer attention to decide which enzyme system is involved. The oxidation of phenols played a major part in the detection of a possible infective agent in human myocardium (Braimbridge *et al.*, 1967; also unpublished data); the formation of adrenochrome-like pigment was significant in studies which re-evaluated the function of mast cells (Chayen *et al.*, 1966c); in each of these studies, the exact oxidative mechanisms were less important than the fact of oxidative activity with respect to phenols and catecholamines.

Rationale. Phenolic compounds, such as dihydroxyphenylalanine (DOPA), tyrosine, adrenaline and catechol or histamine are given to sections of the tissue. If they are oxidized, a coloured quinone, or an adrenochrome-like compound, will form in the section and will be visible because of its inherent colour. Occasionally it may be helpful to add a hydrogen-acceptor like neo-tetrazolium which, acting at an electrode potential close to that of oxygen, may enhance the oxidative effect; the result will still be a melanin even though formazan will also be precipitated (by reduction of the neotetrazolium).

METHOD FOR DOPA-OXIDASE (Diengdoh, 1966a)

Use fresh cryostat sections.

1. Incubate for at least 2 hr in the reaction medium at 37°C.
2. Wash in distilled water.
3. Mount in Farrants' medium.

Control. Before incubation, immerse for 30 min in 10^{-3}M potassium cyanide (3·25 mg/50 ml) in 0·1M phosphate buffer at pH 7·4 at 37°C. Then proceed to the normal incubation in a reaction medium to which 10^{-3}M potassium cyanide has been added.

Result. The test should yield a brown-black pigment which should be lacking in the control.

Reaction medium

0·0056M dihydroxyphenylalanine in 0·1M phosphate buffer at pH 7·4.

[*Note:* at higher pH values, auto-oxidation of the DOPA occurs rapidly. The apparently specific reaction was found to be markedly depressed at pH 6·8.]

Variants. Similar molarities of catechol, or other phenolic compounds may be used in place of DOPA. Also neotetrazolium at a concentration of 0·25 mg/ml may be added together with the substrate to the reaction medium.

MONOAMINE OXIDASE

BIOCHEMICAL DATA

This flavoprotein enzyme (West and Todd, 1961, p. 849) acts on short-chain monoamines; on very long-chain diamines and on tyramine, adrenaline and 5-hydroxytryptamine. It does not act on histamine for which a pyridoxal phosphate enzyme, histaminase, is necessary.

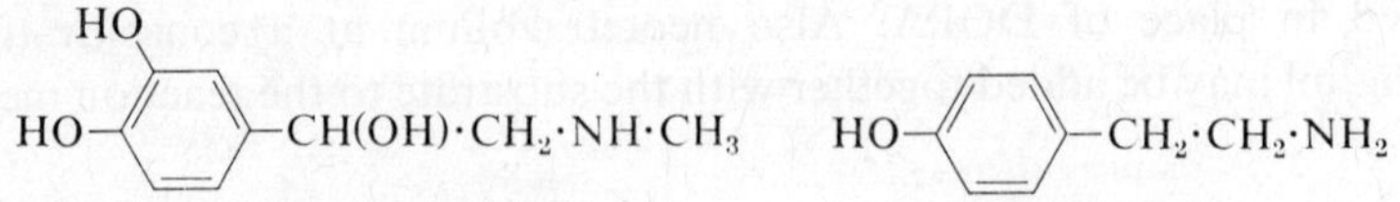

HO
HO—⟨ring⟩—CH(OH)·CH_2·NH·CH_3

adrenaline

HO—⟨ring⟩—CH_2·CH_2·NH_2

tyramine

CH_2·CH_2·NH_2
5
N
H

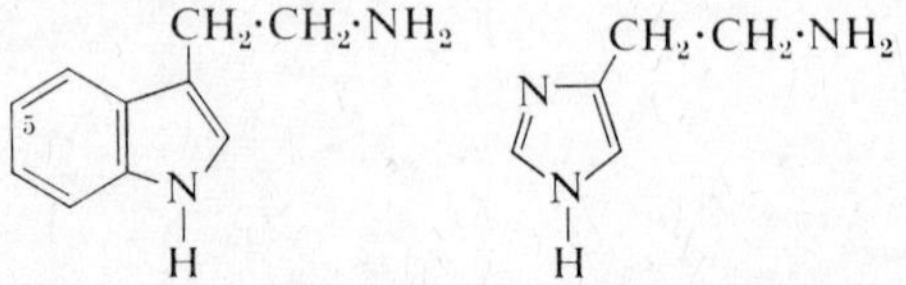

tryptamine
(5-hydroxytryptamine has an
–OH at the position marked 5)

CH_2·CH_2·NH_2
N
N
H

histamine
(not oxidized by
monoamine oxidase)

It is present as an 'insoluble' enzyme in the mitochondrial fraction of homogenates and there seems to be some confusion over the exact chemistry of the enzyme (compare Dixon and Webb, 1964, as against Blaschko, 1961). Moreover there is a possibility that its activity in life may be different from that found when examined manometrically (Blaschko, 1952). Biochemically the reaction is said to occur as follows:

$$R{-}CH_2\overset{+}{N}HR_1 + H_2O + O_2 \longrightarrow R.CHO + \overset{+}{N}H_2R_1 + H_2O_2 \qquad (1)$$

although Blaschko (1952) has suggested that, in life, the immediate hydrogen-acceptor may not be oxygen but some intermediate.

It is inhibited strongly by *iso*propylhydrazine. Ephedrine and other derivatives of phenyl-*iso*propylamine are competitive inhibitors. Cyanide does not inhibit monoamine oxidase (Blaschko, 1952; but see Dixon and Webb, 1964) and this distinguishes this enzyme from other amine oxidases. Marsilid and atabrine have been reported to inhibit monoamine oxidase (Burstone, 1962, p. 456).

$C_6H_5-CH(OH)-CH(CH_3)-NH-CH_3$

ephedrine

$C_6H_5-CH_2-CH(NH_2)-CH_3$

amphetamine (benzedrine)

$C_5H_4N-CO-NH-NH-CH(CH_3)_2$

iproniazid (marsilid)

Monoamine oxidase is found in most tissues. Its highest concentration is found in liver, kidney, intestine and pancreas (Blaschko, 1961).

HISTOCHEMICAL DATA

Blaschko (1952) found that monoamine oxidase activity could be demonstrated by the reduction of triphenyltetrazolium chloride under anaerobic conditions when, apparently, the tetrazole took the place of oxygen in equation (1). He noted that the characteristics of this reaction were different from those found in his biochemical assay procedure. A great deal of work has been done on the mechanism by which it is possible to demonstrate this enzymic activity by the reduction of tetrazoles (see Glenner *et al.*, 1960; Burstone, 1962). In Glenner's scheme, the product of monoamine oxidase activity is indolyl-3-acetaldehyde which spontaneously reduces nitro-blue tetrazolium. This is an interesting idea but it would exclude tyramine and adrenaline from acting as suitable substrates for this method. In fact both can be used; adrenaline is a better substrate than tryptamine or hydroxytryptamine in many tissues. This does not necessarily disprove Glenner's scheme because the whole mechanism of that oxidative activity which histochemists call 'monoamine oxidase' is rather complex. However, we prefer a simpler concept, which is summarized as follows (Cohen *et al.*, 1965):

$$R-CH_2.\overset{+}{N}HR_1+A \xrightarrow[\text{monoamine oxidase}]{-2H} R-CH{:}\overset{+}{N}R_1+AH_2 \tag{2}$$

$$R-CH{:}\overset{+}{N}R_1+H_2O \longrightarrow R-CHO+\overset{+}{N}H_2R_1 \tag{3}$$

$$AH_2+O_2 \longrightarrow A+H_2O_2 \tag{4}$$

It will be seen that the sum total of reactions (2), (3) and (4) is identical with (1). The difference is that this mechanism involves a hydrogen-acceptor, A, which, in the reduced form, can reduce nitro-blue tetrazolium (NBT) to yield a coloured formazan, i.e. reaction (4) can be altered to reaction (5) under anaerobic conditions.

$$R{-}CH_2\overset{+}{N}HR_1 + H_2O + O_2 \longrightarrow R{-}CHO + \overset{+}{N}H_2R_1 + H_2O_2$$

$$+H_2: A \rightarrow AH_2; \quad AH_2 + NBT \rightarrow A + \text{formazan} \qquad (5)$$

As a result of studies on monoamine oxidase in the human endometrium (Cohen *et al.*, 1965) and in various animal tissues (Kirkby, 1965a) it was postulated that this enzyme functions physiologically only inside the mitochondria-like particles but that the onset of menstruation, mechanical damage or histamine can disturb its location, and hence its function, in the affected cells. The disturbed function could be due either to disjunction of the oxidase from its physiological hydrogen-acceptor or to the oxidase being removed from a catalase-like enzyme (which is essential if the peroxide, formed in equation (1), is to be removed; otherwise it will inactivate the monoamine oxidase).

It has also been suggested (Chayen and Bitensky, 1968) that monoamine oxidase plays a major part in the regulation of cellular metabolism.

It should be noted that different tissues metabolize the various substrates for monoamine oxidase at remarkably different rates. Thus it is often necessary to do a preliminary study to decide which substrate should be used. In many of our own investigations, we test the ability of the sections to oxidize tryptamine and also adrenaline and we refer to 'tryptamine (or 5-hydroxytryptamine) oxidation' and 'adrenaline oxidation' separately rather than to 'monoamine oxidase'. In our experience some inhibitors affect each oxidation differently.

METHOD (see Diengdoh, 1966b; Cohen *et al.*, 1965)

Fresh cryostat sections should be used.

1. Incubate in the reaction medium at 37°C for 30 min for adrenaline oxidation and 2 hr for tryptamine or 5-hydroxytryptamine.
2. Wash in distilled water and mount in Farrants' medium.

Result. Monoamine oxidase activity is seen as blue or purple stain, often present in discrete particles.

Controls. The addition of amphetamine (0·01M) to the reaction medium is probably the best control, as this should completely inhibit the reaction. Pre-incubation for 15 min in the complete medium containing amphetamine but lacking the substrate may be necessary, however, for total inhibition.

The use of 0·01M Marsilid (Iproniazid) both for pre-incubation and then to be included in the normal reaction medium has been claimed to give total inhibition in many tissues. We have some evidence, however, that this inhibitor may be metabolized by damaged tissues to a non-inhibitory compound.

Solutions required for this method

Nitro-blue tetrazolium (1 mg/ml) in 0·1M

Phosphate buffer, pH 7·5	5 ml
0·1M phosphate buffer, pH 7·5	5 ml
Distilled water	10 ml

Add substrate at a final concentration of 0·005M [e.g. add 0·02 g of tryptamine hydrochloride].

[*Note:* adrenaline, 5-hydroxytryptamine and Marsilid all cause the pH to drop severely. Because the enzyme seems to be only weakly active outside the pH range 7–8, it is necessary to adjust the pH of the final reaction medium. The difficulty is that alkali, like sodium hydroxide, causes spontaneous reduction of the nitro-blue tetrazolium. Hence for these compounds the buffer should be prepared at pH 8; addition of these substances will then lower the pH of the final reaction medium to the required pH 7·5, but this should be checked.]

PEROXIDASES

Originally Straus used the conventional benzidine histochemical method very successfully for demonstrating horseradish peroxidase in rat tissues after he had injected this purified enzyme intravenously into the animals. But his histochemical demonstration was made possible by the fact that this foreign enzyme was localized in exceptionally high concentration in minute lysosomes (see Straus, 1959, 1962). However Wachstein and Meisel (1964), using a method based on that of Straus, were able to demonstrate endogenous peroxidase activity in sections of a variety of tissues. In plants, Jensen's (1955) guiacol method seems to be useful. Some of the problems connected with reactions for peroxidases have been discussed by van Duijn and his collaborators (e.g. van der Ploeg and van Duijn, 1964a and b). The newer diaminobenzidine method of Graham and Karnovsky (1966) for oxidases and peroxidases may be capable of distinguishing peroxidases separately (Seligman *et al.*, 1968).

DEHYDROGENASES

The dehydrogenases are enzymes which oxidize their substrate by removing hydrogen from the substrate and passing it to a suitable acceptor. They differ from oxidases in that the 'suitable acceptor' cannot be atmospheric oxygen; in almost all cases it is a co-enzyme like nicotinamide-adenine dinucleotide (NAD, or 'DPN' in the older nomenclature), nicotinamide-adenine dinucleotide phosphate (NADP, or 'TPN' in the now discarded nomenclature) or a flavoprotein. The hydrogen is then passed either through a hydrogen-transport system to react ultimately with oxygen, or to an enzymatic process in which it is used for biosynthetic reactions. Either way it is rarely possible to demonstrate the initial dehydrogenase reaction, i.e.

$$AH_2 + NAD \xrightarrow[\text{dehydrogenase}]{} A + NADH_2 \quad (1)$$

(where AH_2 is the substrate and A is the oxidized substrate).

In some cases it may be possible to demonstrate the next stage of this conjoint dehydrogenase–hydrogen-transport mechanism:

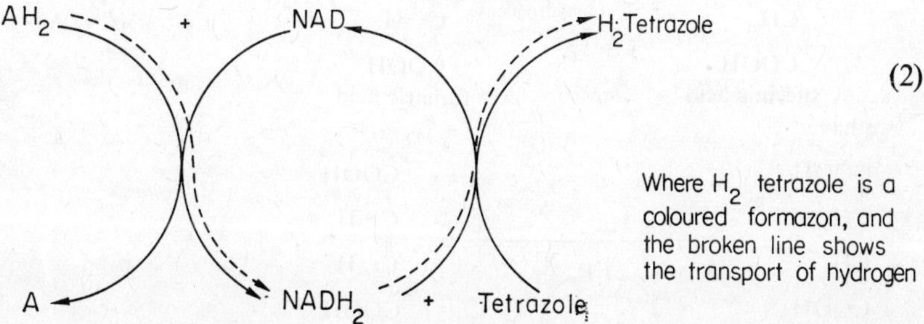

(2)

This need for a hydrogen-transport system has given rise to some confusion in the more theoretical histochemical literature, and some authors (e.g. Burstone, 1962, p. 497) have doubted the value of the histochemical demonstration of dehydrogenase enzymes (also see Farber and Morrison, 1964; Jones, 1965; Farber *et al.*, 1956a and b). Hence some clarification of this problem must be given, even if only at a rather elementary and over-simplified level.

HYDROGEN TRANSPORT AND ELECTRODE POTENTIALS

Hydrogen is removed from the substrate by the appropriate dehydrogenase and passed directly to a hydrogen acceptor (as in (1), above). Thus the substrate is oxidated at the expense of the hydrogen acceptor, which becomes reduced. This, in turn, acts as the hydrogen-donor (in reaction (2)) where $NADH_2$ becomes oxidized back to NAD at the expense of the tetrazole, which becomes reduced to the formazan. So we have a continuous process in which NAD becomes successively and cyclically reduced to $NADH_2$ and then oxidized back to NAD and hydrogen passes from the substrate (shown here as AH_2) to a tetrazole which can take the place of molecular oxygen. Hence we have an oxidation–reduction system, and a hydrogen-transport system.

To the non-specialist, the situation has been complicated by calling this transport of hydrogen 'electron-transport'. This terminology has become widely used but, according to Dixon and Webb (1964, p. 262), the reasons for talking of 'electron-transport' rather than 'hydrogen-transport' are not clear, nor are they based on adequate evidence; they considered the 'electron' terminology to be misleading and would not use it.

As is shown in Fig. 21, the hydrogen moves along the mitochondrial hydrogen-transport system from a relatively high negative electrode potential to the relatively high positive electrode potential of oxygen, at which potential it reacts to form water. So, in the place of the simple equation

$$\underset{\text{succinic acid}}{\begin{array}{c}COOH\\|\\CH_2\\|\\CH_2\\|\\COOH\end{array}} + \tfrac{1}{2}O_2 \xrightarrow[\text{dehydrogenase}]{\text{succinate}} \underset{\text{fumaric acid}}{\begin{array}{c}COOH\\|\\C-H\\\|\\C-H\\|\\COOH\end{array}} + H_2O$$

we have

$$\begin{array}{c}COOH\\|\\CH_2\\|\\CH_2\\|\\COOH\end{array} \xrightarrow[\overset{+}{FP} \rightleftarrows FPH_2]{} \begin{array}{c}COOH\\|\\C-H\\\|\\C-H\\|\\COOH\end{array}$$

$$FPH_2 \dashrightarrow H_2O$$

hydrogen transport system of Fig. 21

where FP is the flavoprotein of succinate dehydrogenase.

The significance of this transport of hydrogen lies in the way energy is released from substrates by oxidation. It could have been thought that oxidation of a substrate (AH_2) itself yields energy. In common parlance, 'you get energy by burning sugar'. But as has been emphasized with great clarity by Krebs and Kornberg (1957), this first step, namely

$$AH_2 + NAD \rightleftharpoons A + NADH_2$$

or

$$BH_2 + NADP \rightleftharpoons B + NADPH_2$$

sets free no appreciable amount of energy (the exception is the oxidation of α-ketonic acids). The hydrogen remains at the same electrode potential; energy is released as it moves down the hydrogen-transport chain; major changes in the free-energy of the oxidation–reduction systems (corresponding

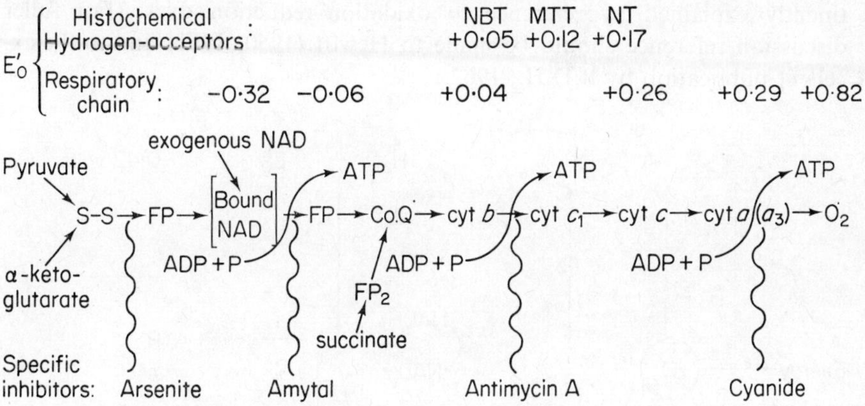

Fig. 21. The hydrogen-transport system of mitochondria.

to major changes in the electrode potential of these systems) occur at certain points in the hydrogen-transport system; the energy liberated by these changes is used for the production of ATP from ADP and inorganic phosphate. This linkage of energy from the oxidative system to provide the energy-rich phosphate bond of ATP is known as oxidative phosphorylation. So, to revert to common parlance, in a sense it is the hydrogen which carries the energy from the sugar; it begins at a high potential energy which it loses during its passage down the hydrogen-transport chain in the mitochondria. In some ways, it is comparable to the state of water in a water-fall, which is being used to drive a turbine (Fig. 22). In the case of a water-fall, the potential energy is converted into kinetic energy as it falls on to each successive ledge (A to B to C to O); in the case of hydrogen-transport it is the change in electrode potential which is being converted into the chemical bond energy of ATP.

The reactions indicated by Krebs and Kornberg (1957) are as follow (NADP can be substituted for NAD):

$NADH_2+FP \rightarrow FPH_2+NAD$ (standard free energy change of -11 kg cal)
FPH_2+oxidized cytochrome$\rightarrow FP+$reduced cytochrome (standard free energy change -16 kg cal)
Reduced cytochrome$+$oxygen$\rightarrow$oxidized cytochrome$+H_2O$ (standard free energy change -25 kg cal)

Total reaction: $NADH_2+\frac{1}{2}O_2 \rightarrow NAD+H_2O$ (standard free energy change -52 kg cal)

One other concept requires to be discussed, and this is the meaning of *electrode potential* and how it affects tetrazoles. It is most simply and pertinently explained by reference to oxidation–reduction dyes. (For fuller discussion reference should be made to Hewitt (1950, 1961) and to the excellent publication by B.D.H., 1962.)

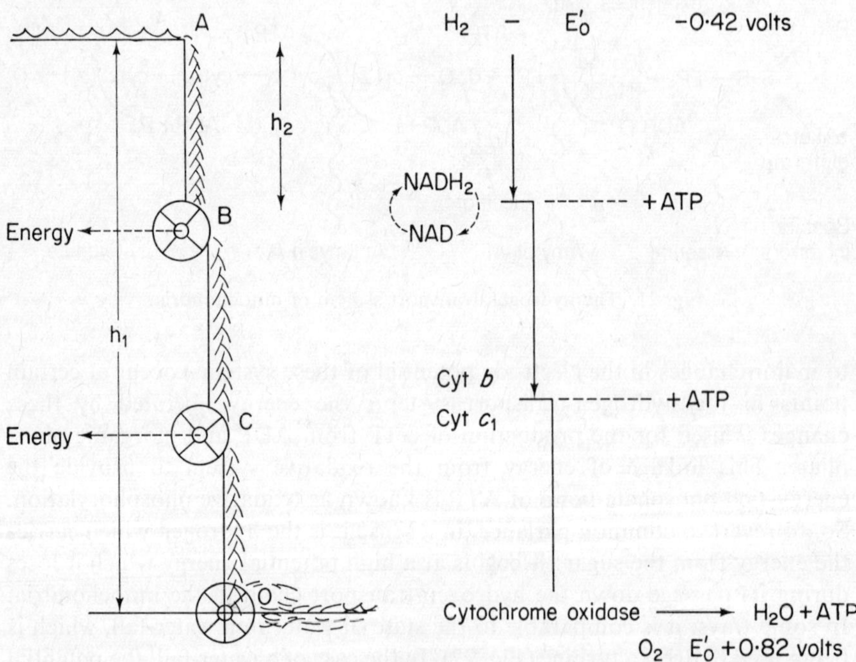

Fig. 22. Analogy between the energy liberated by a water-fall and by the hydrogen-transport mechanism. Potential energy of a mass (m) of water at point A $= mgh_1$ when g is the acceleration due to gravity. Potential energy of the water at B $= mg(h_1-h_2)$ i.e. the energy which is liberated at B $= mgh_2$. At 0 all the potential energy has been finally liberated.

Very many dyes are known which change colour when they become oxidized or reduced; when the change is reversible they are called oxidation–reduction (or redox) indicators. Some which are in common use in biology and medicine are listed in Table 6. If an inert electrode is immersed in a solution of such an indicator which is in equilibrium between its oxidized and its reduced state, the electrode achieves a potential E. This potential (normally expressed relative to that of a normal hydrogen electrode) is directly related to an electrode potential which characterizes that dye (E_0) by a simple equation

$$E = E_0 + 0{\cdot}058 \log \frac{[\text{oxidized dye}]}{[\text{reduced dye}]}$$

where the brackets indicate 'concentration of' the component inside the bracket.

TABLE 6

Some useful redox indicators

Indicator	E_0' *at pH* 7·0
o-Bromophenol-indophenol	+0·230
o-Cresol-indophenol	+0·191
Guaiacol-indo-2:6-dibromophenol	+0·159
Janus green	−0·255
Thionine (Lauth's violet)	+0·063
Methylene blue	+0·011
Methyl viologen	−0·440
Neutral red	−0·325
Phenol-indo-2:6-dichlorophenol (for Vitamin C)	+0·217
Resazurin	Variable
Thymol-indophenol	+0·174

It is seen from this equation that the actual electrode potential measured (E) depends on the relative concentrations of the oxidized and reduced dye. When there is as much of the oxidized as of the reduced form, the equation becomes

$$E = E_0 + 0{\cdot}058 \log 1$$

or

$$E = E_0 \quad (\text{since} \log 1 = 0).$$

In practice the electrode potential which is characteristic for a given oxidation–reduction system is not expressed by E_0 because this is limited, on theoretical grounds, to a constant reaction of pH = 0; consequently it is expressed instead as E_0' which is the normal electrode potential of the system (when the concentration of the oxidized form and of the reduced form are equal, and at any pH other than pH of 0).

The significance of these discussions is as follows: suppose you have two oxidation–reduction systems of different E_0', the one $X \rightleftharpoons XH_2$ having an E_0' of $+0{\cdot}1$ volt and the other $Y \rightleftharpoons YH_2$ with an E_0' of $-0{\cdot}1$ volt. Then it follows that the first system will be able to oxidize the second, i.e. $YH_2 + X \rightarrow Y + XH_2$. The general rule is that oxidizing systems have the more positive potentials; reducing systems have the more negative potentials; or, in other words, the more negative the electrode potential, the greater is its reducing power. (Other conventions do exist but this seems to be the most generally used, e.g. see Burton, 1961; Hewitt, 1961; West and Todd, 1961, p. 821.)

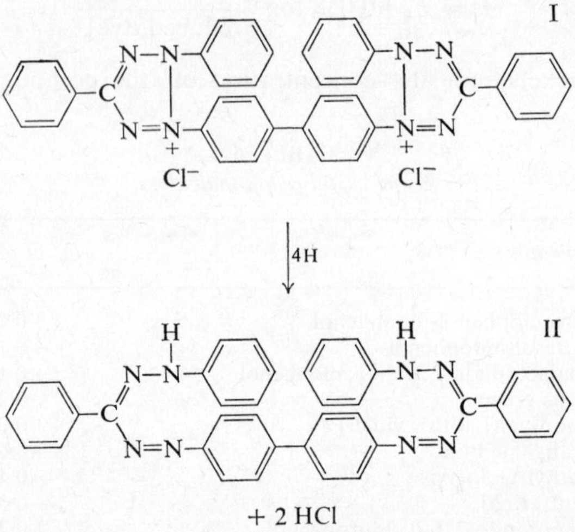

I

II

+ 2 HCl

Fig. 23. Formulae of neotetrazolium chloride (I) and its corresponding formazan (II).

But the point which deserves to be stressed is that this reaction of $YH_2 + X \rightarrow Y + XH_2$ does not require any enzymatic activity. Thus the oxidation–reduction potential of a system can be estimated by adding to it various redox indicators of known E_0' and finding which are affected. This method has been used for studying the potentials of living tissues. In the example quoted by the B.D.H. booklet, if a tissue reduces an indicator of $E_0' - 0{\cdot}05$ volt but does not reduce one of $E_0' - 0{\cdot}15$ volt, the oxidation–reduction potential of the tissue must lie between $-0{\cdot}15$ and $-0{\cdot}05$ volts.

Similarly the hydrogen passing down the hydrogen-transport chain in mitochondria will reduce an oxidation–reduction indicator when its latest acceptor system has an electrode potential which approximates to (but is still more negative than) that of the indicator. This will be true of any chain of oxidation–reduction systems whether they are present in mitochondria or not; emphasis is placed on mitochondrial systems only because they are better (but still incompletely) understood than are the comparable systems of microsomes or of the rest of the cytoplasm. The tetrazolium salts, which become reduced at particular electrode potentials to form insoluble formazans, come within this scheme (see Fig. 21). Consequently when dehydrogenase activity is demonstrated histochemically the requisite substrate must be given in the presence of a tetrazole: hydrogen is removed from it by the enzyme and is passed to one or more hydrogen acceptors until it reaches an oxidation–reduction system which has an electrode potential which is suitable for reducing the tetrazolium salt. It has become customary to refer to this enzyme not by its biochemical name alone but by reference also to the tetrazole. For example, the enzyme succinate dehydrogenase, when demonstrated histochemically by the reduction of neotetrazolium in the presence of succinate, is often called succinate-neotetrazolium reductase (e.g. see Novikoff, 1963b). To us this seems unnecessary; on occasion this nomenclature may even be undesirable because it confuses the mechanism of the reaction and makes it more difficult to relate the histochemical response to the biochemistry of the metabolism of the tissue.

GENERAL HISTOCHEMISTRY OF DEHYDROGENASES

As has been discussed (immediately above, and in the Introduction to Enzyme Histochemistry, p. 100) fresh cryostat sections are incubated at 37°C in the presence of (*i*) the substrate which is to be oxidized by the dehydrogenase; (*ii*) any co-factor that is required as the immediate hydrogen-acceptor of the dehydrogenase activity (e.g. NAD or NADP); (*iii*) any activators of the enzyme that may be known from biochemical or histochemical studies; and (*iv*) a tetrazolium salt which acts as the trapping agent for the hydrogen and, in becoming reduced by it, yields a coloured precipitate of formazan (Fig. 23). We have seen that there may be a considerable gap between the electrode potential at which the co-factor becomes reduced and that at which the tetrazole can accept hydrogen. This gap (Fig. 21) is greater for neotetrazolium (NT) than it is for nitro-blue tetrazolium (NBT). The newer evidence of Altman (1972) suggests that it is even less for MTT and for INT. But with NT or NBT the hydrogen must be carried by a hydrogen-transport system such as that of mitochondria (Fig. 21) or of the microsomal respiratory chain (Fig. 28) of the cytoplasm until it can react with the tetrazole. This involves two features: firstly, the localization of the formazan at best is that of the part of the hydrogen-transport chain from which it accepts the hydrogen; secondly, the rate at which the dehydrogenase works may be considerably greater than the rate at which the hydrogen-transport chain passes the hydrogen so that the amount of dehydrogenation which will be detected will depend on the activity of the hydrogen-transport system and not of the dehydrogenase. At least as far as the cytoplasmic (non-mitochondrial) dehydrogenases are concerned, it seems that these problems can be overcome by including phenazine methosulphate (PMS) in the reaction medium (usually at a concentration of 0·2 mg per ml). This substance is a powerful hydrogen-acceptor which very rapidly and non-enzymatically takes hydrogen from NADPH, NADH and from the reduced flavo-protein of succinate dehydrogenase (Singer and Kearney, 1954) and transmits it to tetrazoles, including NT (Farber and Bueding, 1956; Chayen *et al.*, 1966a; for general review see Altman, 1972). Consequently the histochemist is in the powerful position where (by using PMS) he can detect, and measure, the amount of hydrogen passed on to the cofactor by the dehydrogenase and (by leaving out the PMS) he can determine how well the hydrogen-transport system transmits this hydrogen. By using different tetrazolium salts, which become reduced at different electrode potentials, he can test each level of the hydrogen-transport chain. These technical advances have opened new aspects of cellular metabolism in disease and in drug action (e.g. Chayen *et al.*, 1972). Moreover the use of colloid stabilizers, such as polyvinyl alcohol (see Altman, 1972, for review of this subject) or collagen polypeptides (Butcher, 1971b) together with PMS may allow more precise localization of the initial dehydrogenation site.

The same general arguments apply to mitochondrial dehydrogenases except that the issue is slightly complicated by the relative impermeability of the intact mitochondrial membrane. Colloid stabilizers also protect these membranes.

GENERAL METHODS (also see Nachlas *et al.*, 1958a and b; Farber *et al.*, 1956a and b; Sternberg *et al.*, 1956)

To test whether a section can oxidize any given substrate is a relatively simple task. It is obvious that optimal concentrations of substrate, co-factor, activators, and hydrogen-acceptor should be used if possible, particularly because examples have been described where dehydrogenases have been partially or even completely inhibited by excess of substrate, of activator or of tetrazolium salt in the usual histochemical procedures. But for dehydrogenases which are not specifically described in this section, it is possible to obtain some idea of the activity present in the section by incubating for ½–1 hr at 37°C in a medium containing the substrate at a concentration of 0·05M, NAD (where required) at a concentration of 5 mg per ml and nitro-blue tetrazolium (0·1% solution) in a phosphate or glycyl glycine buffer (0·1M) at a pH of 7·5. If the dehydrogenase requires NADP, the substrate concentration should be 0·005M and that of the NADP should be 1 mg per ml. Neotetrazolium can be used in place of nitro-blue tetrazolium at the same concentration (0·1%); the incubation time will then be about 2 hr.

To avoid having the tetrazolium salt and atmospheric oxygen both competing for the hydrogen and for other reasons, including the avoidance of deleterious free radicals during the incubation (e.g. see Tappel *et al.*, 1963; Desai and Tappel, 1963; Slater, 1966), it is advisable to do such incubation in an atmosphere of nitrogen (see p. 10).

RESULTS IN DEHYDROGENASE HISTOCHEMISTRY

As has been discussed immediately above (also see Fig. 21), the tetrazole traps the hydrogen when the latter reaches an oxidation–reduction system of electrode potential compatible with the tetrazole which is used. The localization of the formazan produced by this reduction of the tetrazole depends upon the following factors:

1. The rate of dehydrogenase activity. As discussed for enzyme histochemistry generally (p. 101), the insoluble end-product—in this case the formazan—is truly insoluble only if its local concentration exceeds a certain value. Consequently, if the enzymic dehydrogenation is slow, or if the transport of the hydrogen is slow, the rate at which the formazan is produced may be too slow to ensure that it will be precipitated at the site at which it becomes reduced. It can then diffuse to other sites in the tissue (e.g. to fat droplets). For this reason it is inadvisable to use chemical fixation which will denature the dehydrogenase and so reduce its activity.

2. The 'substantive' properties of the formazan. Certain tetrazoles, such as neotetrazolium, tend to crystallize. Hence their localization will depend on where they can find centres for crystallization; if fatty matter is in the vicinity, they may dissolve in this before they have the opportunity to crystallize. The formazan of nitro-blue tetrazolium, on the other hand, does not crystallize but appears to be a good stain for tissue, i.e. as it forms, so it tends to stick to the nearest tissue element (see Nachlas *et al.*, 1957b). Consequently it may be expected to give a good indication of the site at which it accepted the hydrogen. (Other tetrazolium salts exist, each with its own electrode potential, efficiency of hydrogen-acceptance and 'substantive' properties. They must be evaluated separately by the worker who wishes to use them.)

3. The site of hydrogen-uptake. We have emphasized that tetrazolium salts accept hydrogen at a particular electrode potential; they cannot accept it from the dehydrogenase enzyme itself. This point has been strongly stressed by Farber *et al.* (1956a, b; Sternberg *et al.*, 1956; also see Nachlas *et al.*, 1958a, b). Its implications deserve some consideration.

(*i*) *Enzymes bound in mitochondria.* An enzyme, such as succinate dehydrogenase, which is tightly bound to the mitochondrial structure is, in fact, in close proximity to the whole hydrogen-transport system. Consequently it is likely that the deposition of nitro-blue formazan will indicate the true localization of such an enzyme, within the limits of resolution of the light microscope, irrespective of which part of the hydrogen-transport system acts as the hydrogen-donor.

(*ii*) *Dehydrogenases bound in the cytoplasm.* If the dehydrogenase remains bound in the cytoplasm even during the histochemical incubation procedure,

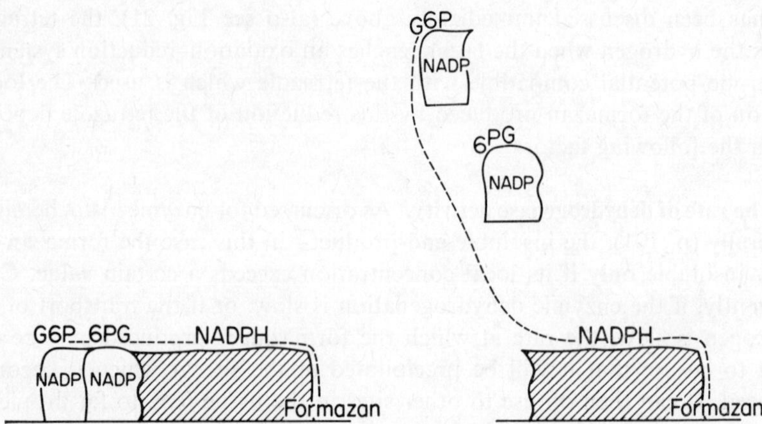

Fig. 24. A diagrammatic representation of what occurs during incubation for glucose-6-phosphate and 6-phosphogluconate dehydrogenases in the presence (on the left) and absence of PVA. The enzymes are shown as small bags with NADP and the relevant substrate (G6P for the former and 6PG for the latter) attached to them. In the presence of PVA, the dehydrogenases are linked together and to their diaphorase system (shown cross-hatched). The reduced NADP is immediately converted to the formazan. In the absence of PVA (on the right) the enzymes become dispersed into the reaction medium and bring about the reduction of the NADP in free solution. The reduced NADP (NADPH) then acts as a soluble substrate for the tissue-bound diaphorase. (Figure by courtesy of Dr. F. P. Altman.)

it will liberate the reduced co-enzyme at its own site. In studies on the pentose-shunt dehydrogenases (Altman and Chayen, 1966a), it has been shown that the dehydrogenases are actually linked by intermolecular linkage to the relevant diaphorase (hydrogen-transport to oxygen) system. Consequently an efficient tetrazole like nitro-blue tetrazolium will demonstrate the actual site of the dehydrogenase.

(*iii*) *Non-bound dehydrogenases.* In a number of circumstances, dehydrogenases may not be bound to the tissue and will then act on their substrate to oxidize it at the expense of soluble NAD (or NADP), producing $NADH_2$ (or $NADPH_2$) in the incubation medium. This reduced co-enzyme is then the soluble substrate for the equivalent diaphorase system and the tetrazole, in accepting hydrogen from the diaphorase hydrogen-transport mechanism, locates that system rather than the primary dehydrogenase. In many histochemical procedures for the demonstration of dehydrogenase activity, this is a real problem. The reason is that most incubation techniques are highly disruptive and liberate the dehydrogenase from its normal location in the cell; in some cases this can be completely overcome by the use of a suitable concentration of polyvinyl alcohol (PVA) in the incubation medium (see pp. 11–14).

THE HISTOCHEMISTRY OF SELECTED DEHYDROGENASES

SUCCINATE DEHYDROGENASE

BIOCHEMICAL DATA

The enzyme is an essential part of Krebs' tricarboxylic acid cycle (Fig. 25a and b) and so is found in all aerobic cells. It is fairly firmly bound to the

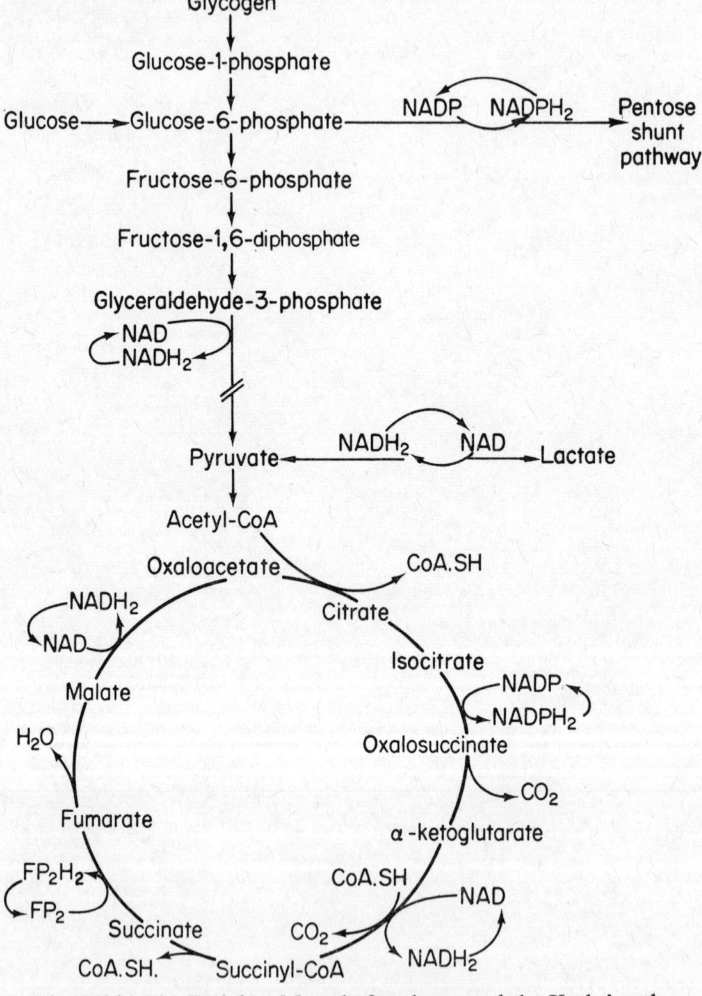

Fig. 25(a). The Embden–Meyerhof pathway and the Krebs' cycle.

framework of the mitochondria but has been isolated and purified; it is then very unstable. The isolated enzyme contains a flavine prosthetic group which

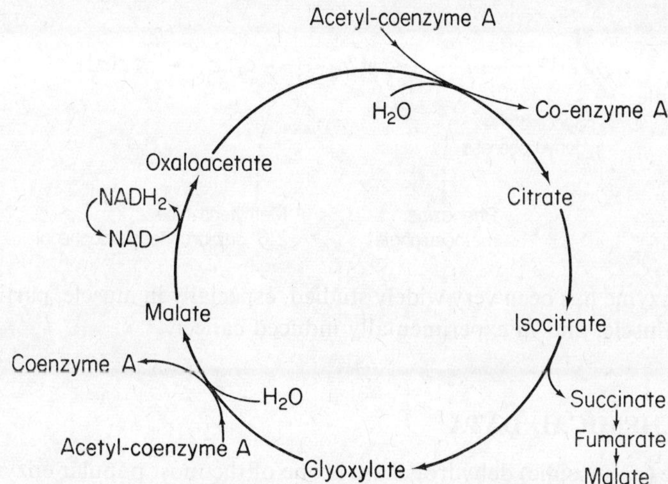

Fig. 25(b). The glyoxylate cycle. Although this involves some of the Krebs' cycle enzymes, succinate dehydrogenase is not directly involved.

is related to FAD (flavine–adenine dinucleotide); its pH optimum is about pH 7·6.

Succinate dehydrogenase has many competitive inhibitors. The most significant in histochemical studies is malonate which has the two carboxyl groups with approximately the same spacing between them as has succinate but which lacks the oxidizable $-CH_2-CH_2-$ lying between them (see Dixon and Webb, 1964, p. 352). Other inhibitors include phosphate, fluoride, suramin and oxaloacetate (*Biochemists' Handbook*, p. 370). It is also inhibited by reagents which combine with $-SH$ groups.

The oxidation of succinate is as follows:

$$\begin{array}{lcl} \text{COOH} & & \text{COOH} \\ | & & | \\ \text{CH}_2 \quad +\text{A} & \rightleftharpoons & \text{CH} \quad +\text{AH}_2 \\ | & & \| \\ \text{CH}_2 & & \text{CH} \\ | & & | \\ \text{COOH} & & \text{COOH} \\ \text{succinic acid} & & \text{fumaric acid} \end{array}$$

where A is a hydrogen-acceptor.

According to Slater (1961) the hydrogen is taken away from the succinate along the hydrogen-transport system which can be 'short circuited' by the use of suitable hydrogen-acceptors, as shown below.

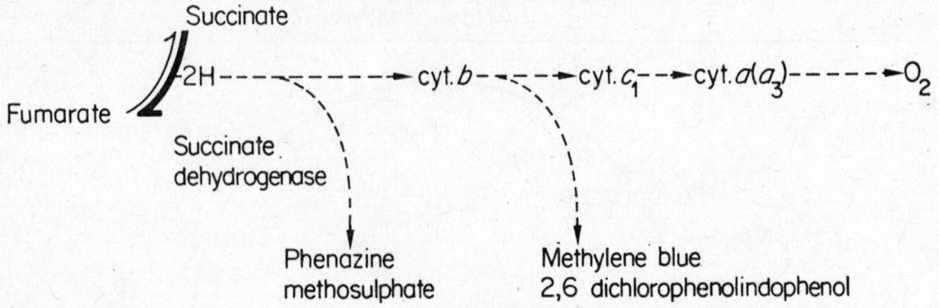

Succinate

Fumarate ⇌ 2H - - - → cyt. b - - - → cyt. c_1 - - → cyt. $a(a_3)$ - - - - → O_2

Succinate dehydrogenase

Phenazine methosulphate

Methylene blue
2,6 dichlorophenolindophenol

This enzyme has been very widely studied, especially in muscle, particularly cardiac muscle, and in experimentally induced cancer.

HISTOCHEMICAL DATA

Succinate (or succinic) dehydrogenase is one of the most popular enzymes for histochemists who use it as a marker for the Krebs' cycle (Fig. 25). It has been much investigated in cancers and reference should be made to one of the standard textbooks (such as Burstone, 1962) for fuller details. The results have been very varied, some cancers having high and others very low activities. However, when investigated in detail, it has been shown that histochemistry can give special information which cannot be obtained readily by biochemical procedures. For example, there has been a huge amount of work done on the experimental induction of cancer in livers of rats fed with butter yellow (dimethylaminoazobenzene), or some allied compound. It has been shown conclusively, biochemically, that the succinate dehydrogenase activity of the affected liver drops to about 50% of its normal level at the time when the malignant change occurs, and this is related to a reduction in the number of mitochondria (see Potter *et al.*, 1950; Fiala and Fiala, 1959). It should be noted that the biochemical data are related to unit weight of liver. Histochemical analysis showed that this drop in total activity per unit mass was due to proliferation of cells from the bile duct, which have few mitochondria (in the normal sense) and very low succinate dehydrogenase activity even in normal healthy liver. The activity in each hepatic cell is unchanged. But with the considerable proliferation of bile duct cells, there is an increase in the proportion of cells of low succinate dehydrogenase activity in the unit mass of liver, i.e. the apparent decrease in activity of this enzyme is an artefact caused by the dilution of its normal (and unaffected) activity by bile duct cells which have very low activity. This effect was called 'the tissue dilution artefact' (Chayen *et al.*, 1961; also see Jones *et al.*, 1961; Jones, 1963).

Succinate dehydrogenase activity has been studied by quantitative histochemistry (see Part III) by Defendi and Pearson (1955), by Jones *et al.* (1963) and by Chayen *et al.* (1966a). It has also been shown to play a significant part in human myocardial dysfunction (Niles *et al.*, 1964a, 1966).

METHODS

For all methods, fresh cryostat sections of unfixed tissue must be used.

Simple method

1. Incubate in the reaction medium in a Coplin jar, or in a ring, at 37°C in an atmosphere of nitrogen. With neotetrazolium the incubation requires 2 hr; with nitro-blue tetrazolium about ½ hr.
2. Wash well in distilled water.
3. Mount in Farrants' medium.

Result

1. With neotetrazolium: red-blue granules and red colour, which is absent in the control, show amount of succinate dehydrogenase activity.
2. With nitro-blue tetrazolium: red to purple staining, which is not found in the control, indicates both the amount of succinate dehydrogenase activity and the site at which hydrogen, removed from the succinate, has been trapped by the tetrazole.

Control

1. Incubate in a 0·05M solution of sodium malonate in the buffer for 15 min at 37°C.
2. Incubate in the complete reaction medium, to which 0·05M malonate (0·37 g sodium malonate in 50 ml of the reaction medium) has been added.
3. Wash.
4. Mount in Farrants' medium.

This treatment should inhibit all true succinate dehydrogenase activity.

Solutions required for this method

Reaction medium

Add 0·1 g of neotetrazolium chloride (or nitro-blue tetrazolium) to 100 ml of 0·1M phosphate buffer (or 0·05M glycyl glycine) at pH 7·8. Heat gently to dissolve. Cool. Filter. Adjust to pH 7·8 if necessary.

To 50 ml of this solution, add 0·68 g of sodium succinate ($6H_2O$); this gives a 0·05M solution.

Bubble nitrogen through the solution at 37°C before and during use.

Phenazine methosulphate method
Incubate as for the simple method, but add 2 mg of fresh phenazine methosulphate (PMS) to 50 ml of the normal reaction medium. The incubation time can be reduced if required. It is advisable to agitate the solution during the incubation.

Result. This compound (PMS) usually enhances the reaction with either tetrazole. Because it accepts hydrogen from the flavoprotein of the dehydrogenase it should give more precise localization of the enzyme especially if nitro-blue tetrazolium is used as the acceptor. But care must be taken to use fresh PMS because it appears to decompose even when stored dry for a long period.

PVA method (Butcher, 1970)
The addition of a colloid stabilizer, such as polyvinyl alcohol (PVA) or the collagen polypeptide 5115 protects unfixed sections undergoing reaction at relatively physiological pH values (see Part II). Such stabilizers are therefore essential for weakly bound or 'soluble' enzymes. In the study of membrane-bound enzymes, as for those in mitochondria, PVA in particular may so stabilize the membranes that the reaction is reduced because substrate is restricted from entering the organelle. Phosphate buffer may assist penetration, even in such stabilized sections, very much as it does in biochemical studies.

The method is as for non-stabilized sections (above). The reaction medium contains 0·3% of neotetrazolium chloride (or 0·4% nitro-blue tetrazolium) in 0·05M glycylglycine (or 0·1M phosphate) buffer, pH 8·0. To 100 ml of this solution is added 20 g of PVA (see under 'Glucose-6-phosphate dehydrogenase', p. 209), and 50mM of sodium succinate. When required, 0·2 mg/ml of PMS can be included in the reaction medium.

SIMILAR DEHYDROGENASES

The same histochemistry can be applied to any dehydrogenases which do not require an added co-factor, such as NAD. Of these enzymes probably the most important is α-glycerophosphate dehydrogenase. There are two glycerophosphate dehydrogenases and these are involved in the dihydroxyacetone–phosphate shuttle, changes in which have been implicated in the change to malignancy (e.g. Sacktor and Dick, 1960; Boxer, 1965; Stuart *et al.*, 1970; Chayen *et al.*, 1972). The significance of the shuttle is as follows: NAD reduced in the cytoplasm cannot penetrate through the mitochondrial membrane. Cytoplasmic NADH is therefore used to reduce dihydroxyacetone–phosphate to form α-glycerophosphate which passes readily through the mitochondrial membrane and gives up its hydrogen, inside the mitochondrion, under the influence of the mitochondrial glycerophosphate dehydrogenase which does not require NAD as co-factor:

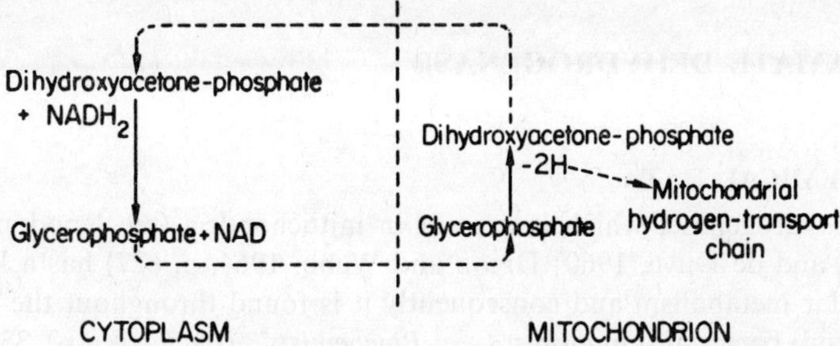

In this way (e.g. see Lehninger, 1965) the reducing equivalents (i.e. hydrogen on NAD) of the cytoplasm are transported by the reduced product, namely α-glycerophosphate, into the mitochondrion where they are liberated to yield ATP energy via the mitochondrial hydrogen-transport chain (by normal oxidative phosphorylation).

In histochemistry it is usually necessary to add either PMS or menadione sodium bisulphite salt ($0{\cdot}6 \times 10^{-3}$M) to the reaction medium to demonstrate adequately the activity of this NAD-independent enzyme.

GLUTAMATE DEHYDROGENASE

BIOCHEMICAL DATA

This dehydrogenase, which is present in mitochondria (see Roodyn, 1965; Bendall and de Duve, 1960; Dixon and Webb, 1964, p. 627) has a key role in cellular metabolism and consequently it is found throughout the animal, plant and bacterial kingdoms (see *Biochemists' Handbook*, p. 333; also Dixon and Webb, 1964). In some tissues it depends on NAD, in others on NADP, causing the following oxidative deamination (West and Todd, 1961, p. 1060):

(*i*) Oxidation:

$$\underset{\text{L-glutamic acid}}{\begin{array}{c}CH_2-CH_2-COOH\\|\\H_2N-CH-COOH\end{array}} + NAD \rightleftharpoons \underset{\alpha\text{-iminoglutaric acid}}{\begin{array}{c}CH_2-CH_2-COOH\\|\\HN{=}C-COOH\end{array}} + NADH_2$$

NADP can replace NAD in this reaction.

(*ii*) Deamination:

$$\underset{\alpha\text{-iminoglutaric acid}}{\begin{array}{c}CH_2-CH_2-COOH\\|\\HN{=}C-COOH\end{array}} + H_2O \xrightarrow[\text{spontaneous reaction}]{} \underset{\alpha\text{-ketoglutaric acid}}{\begin{array}{c}CH_2-CH_2-COOH\\|\\O{=}C-COOH\end{array}} + \underset{\text{ammonia}}{NH_3}$$

In intact mitochondria, this enzyme is relatively inactive, becoming more active as the mitochondria are deliberately disrupted (see Bendall and de Duve, 1960). It is also involved in various transaminase reactions which have become of significance because the levels of serum transaminases are used in clinical biochemistry for the assessment of tissue death in the liver and in the heart (see e.g. Maclagan, 1964; King, 1965). Bitensky (1967b) has suggested that glutamic dehydrogenase activity may be closely linked with transaminase activity in damaged cells by the following type of scheme shown opposite.

In this scheme, increased activity of the glutamate dehydrogenase will be expected to shift the transamination equilibrium from left to right in the normal transamination process,

$$\text{Protein} \rightleftharpoons \text{amino-acid} + \alpha\text{-ketoglutarate} \underset{\text{transaminase}}{\rightleftharpoons} \text{glutamate} + \text{keto-acid}$$

and so enhance the katabolism of protein.

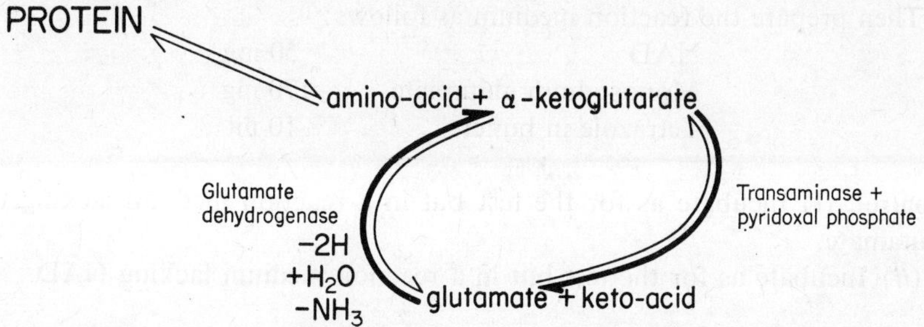

The pH optimum of glutamate dehydrogenase, when oxidizing glutamate, is about pH 8–8·5, and depends on the buffer and the ionic concentrations. The dehydrogenase is inhibited by *p*-chloromercuribenzoate, by Ag^{+}, Hg^{++} and by Fe^{+++}.

HISTOCHEMICAL DATA

This enzyme reaction has been found to be a very sensitive indicator of cellular damage (see Kirkby, 1965b; Bitensky, 1967b; Chayen and Bitensky, 1968). The cause for the increased activity appears to be an increased permeability of mitochondria, in damaged cells, to the substrate and co-enzyme (Chayen *et al.*, 1966b). The enzyme activity can be measured by quantitative histochemistry.

METHOD

Fresh cryostat sections of unfixed tissue must be used. Care must be taken not to damage the tissue sample mechanically: even pressure by forceps can be detected in the final reaction if it occurs sufficiently long before the chilling is done.

1. Incubate in the reaction medium in an atmosphere of nitrogen either in a ring or in a Coplin jar. Incubate for 2 hr at 37°C if neotetrazolium is used; with nitro-blue tetrazolium, 30 min should be sufficient.
2. Wash in distilled water.
3. Mount in Farrants' medium.

Result. Blue, red or purple formazan (depending on the tetrazole used) indicates glutamate dehydrogenase activity if it is not present in the control.

Solutions required for this method

Reaction medium

Dissolve 0·1 g of neotetrazolium chloride, or of nitro-blue tetrazolium in 100 ml of 0·05M phosphate buffer (or 0·05M glycyl glycine) at pH 7·8. Warm to dissolve. Cool. Filter.

Then prepare the reaction medium as follows:

NAD	50 mg
Monosodium glutamate	170 mg
Tetrazole in buffer	10 ml

Control. (*i*) Incubate as for the test but in a reaction medium lacking the glutamate.

(*ii*) Incubate as for the test but in a reaction medium lacking NAD.

LACTATE DEHYDROGENASE

BIOCHEMICAL DATA

The most commonly found lactate dehydrogenase reversibly oxidizes lactate to pyruvate as in the following equation:

$$\begin{array}{c} CH_3 \\ | \\ CHOH \\ | \\ COOH \end{array} + NAD \rightleftharpoons \begin{array}{c} CH_3 \\ | \\ C{=}O \\ | \\ COOH \end{array} + NADH_2$$

lactic acid → pyruvic acid

It is found in all cells which are capable of glycolysis. In the formation of lactic acid in muscle, it operates by reducing pyruvate, i.e. the equation goes from right to left, with optimal activity around pH 7·4; its optimal oxidation of lactate is said to be close to pH 10. In animal tissues it can react with a number of α-hydroxy-acids as well as lactic acid. When isolated from ox heart its molecular weight is about 135,000; from rat liver it is about 126,000. The enzyme from muscle contains zinc (*Biochemists' Handbook*, p. 322). In yeasts the lactate dehydrogenase is not dependent on NAD and is identical with cytochrome b_2 (*Biochemists' Handbook*, p. 323).

In clinical biochemistry this enzyme has proved of some value in indicating which tissue has undergone such trauma as to cause cell death. This comes about because five iso-enzymes of lactate dehydrogenase have been described; the pattern of these, as seen on electrophoresis of the serum, indicates the tissue from which they have been liberated (see Plagemann *et al.*, 1960a and b). For example, the pattern obtained when heart tissue is damaged contains mainly lactate dehydrogenase iso-enzymes 1 and 2; when the liver has been damaged iso-enzyme No. 5 predominates. Cahn *et al.* (1962) have indicated that each molecule of lactate dehydrogenase is made up of four sub-units; each iso-enzyme would then be a hybrid containing various proportions of each of the two sub-units (H and M) so that the five iso-enzymes would be HHHH, HHHM, HHMM, HMMM and MMMM. Thus it would be conceivable that the existence of these five iso-enzymes could be due to the activity of two genes (also see Dixon and Webb, 1964).

Lactate dehydrogenase is present almost exclusively in the soluble fraction of tissue homogenates as a 'soluble' enzyme (Dixon and Webb, 1964, p. 627). It contains —SH groups and can be inhibited by *p*-chloromercuribenzoate.

HISTOCHEMICAL DATA

The equilibrium of the reaction

$$\underset{\text{lactic acid}}{CH_3CHOH.COOH} + NAD \rightleftharpoons \underset{\text{pyruvic acid}}{CH_3.CO.COOH} + NADH_2$$

can be upset to ensure that the reaction goes from left to right. In the histochemical methods used for this reaction, the equilibrium can be upset in the required direction by one of two ways. Generally it is sufficient to allow the $NADH_2$-diaphorase in the tissue to re-oxidize the $NADH_2$ to NAD; the diaphorase hands the hydrogen on to the tetrazole which, precipitating to form the formazan, removes the hydrogen from the equilibrium system; the diaphorase is thus free for further oxidation of newly formed $NADH_2$. The second possibility is to add cyanide to the reaction medium; this reacts with the pyruvate and removes it from the equilibrium. It will be appreciated that (unless phenazine methosulphate is used to short-circuit the diaphorase system) the deposition of formazan will reflect the activity of the combined dehydrogenase-diaphorase system; at best it is liable to show the localization of the diaphorase.

METHOD

Use fresh cryostat sections of unfixed tissue.

1. Incubate in the reaction medium at 37°C in rings (or in a humidity chamber, or in a Coplin jar). The duration of incubation can be decided by inspection of the intensity of colour produced by the formazan. Generally 2 hr is enough when neotetrazolium is used as the hydrogen-acceptor and ½ hr for nitro-blue tetrazolium.
2. Wash in distilled water.
3. Mount in Farrants' medium.

Result. Red, blue, or purple formazan demonstrates lactate dehydrogenase activity.

Control. As the test but excluding first the lactate, and then the NAD from the reaction medium.

Reaction medium

Dissolve 0·2 g of sodium lactate in 25 ml of phosphate-tetrazole solution (as used for succinic dehydrogenase: 1 mg/ml of neotetrazolium or of nitro-blue tetrazolium in 0·05M phosphate buffer at pH 8·0). Check that the pH is still pH 8·0 after dissolving the lactic acid. To 1 ml of this solution add 5 mg of NAD.

PVA METHOD

Because lactate dehydrogenase is a soluble enzyme it is probably best to use a colloid-stabilizer in the reaction medium when demonstrating its activity. McMillan (1967) used a 30% solution of Dextran-10 (average molecular weight 11,200; from Pharmacia Fine Chemicals) and PMS; he inhibited the M-iso-enzyme by 4M urea and the H-iso-enzyme by including pyruvate to give a concentration of lactate : pyruvate of 10 : 1 (see Biddy and Engel, 1964). Polyvinyl alcohol (PVA) has been used by many workers for soluble, NAD-dependent dehydrogenases (Kunze, 1967; Jacobsen, 1969; Wenk, 1970; Dahl and From, 1971).

Reaction medium

Dissolve 50 mM of sodium lactate in a 0·05M glycyl glycine buffer, pH 8·0, to which has been added the tetrazolium salt (5mM of nitro-blue tetrazolium, 0·4%, or of neotetrazolium chloride, 0·3%) and PVA at 20 g to 100 ml of the buffer. (For details, see under 'Glucose-6-phosphate dehydrogenase'). To 1 ml of this solution add 2·5 mg of NAD. Check the pH is 8·0. If required, PMS may be added at a concentration of 0·2 mg/ml. The solution should be saturated with nitrogen before use.

It is particularly important to have a control, which contains all these reactants except lactate.

STEROID DEHYDROGENASES

These enzymes are of great importance in the metabolism of steroids, in the role of steroids as hormones and so in cellular physiology and in endocrinology. Their biochemistry and histochemistry have been discussed fully recently by Baillie *et al.* (1966) who have tended to prepare and incubate their specimens by the methods described in this book. In general, these dehydrogenases are more active with NAD than with NADP, and so are considered together with other NAD-dependent dehydrogenases. The special problems posed by these enzymes are (*i*) their solubility (which may be overcome by the use of PVA, as will be discussed for pentose-shunt dehydrogenases, p. 207; also see Baillie *et al.*, 1966, p. 8); (*ii*) the insolubility of the steroid substrates. This insolubility is overcome by dissolving the steroid substrate (5 mg) in 0·5 ml of dimethyl formamide and dispersing this in the tetrazole-buffer containing NAD. For some steroids, particularly 17β-oestradiol, as little as 50 μg will give good reactions in placenta (for this information we are indebted to Dr. D. M. Hart).

The reaction medium generally (Baillie *et al.*, 1966, p. 8) contains:

2 mg/ml of NAD (or of NADP, but all reactions seem to be greater when NAD is used),

0·5 mg/ml of nitro-blue tetrazolium,

0·25 mg/ml of the selected steroid (dissolved in dimethyl formamide),

all in 0·1M Tris or phosphate buffer at pH 7·5.

The sections (8–15 μ as desired) are incubated for various periods (e.g. 2 hr) at 37°C. The control lacks the substrate but should include the same volume of dimethyl formamide (also see Hart, 1966).

OTHER NAD-DEPENDENT DEHYDROGENASES

We have now considered a mitochondrial dehydrogenase (glutamate), a 'soluble' dehydrogenase (lactate), and the awkward (because of the insolubility of the substrate) steroid dehydrogenases. All of these are NAD-dependent; the methods for demonstrating them histochemically can act as indications for identifying the other NAD-dependent dehydrogenases which may interest histochemists. It must be stressed that these only indicate the histochemical procedures which can be tried. For rigorous study, each enzyme can be expected to have its own requirements for the optimal concentrations of substrate, of NAD and of tetrazole, as well as its own optimal pH.

The NAD-dependent dehydrogenases which may be obviously significant in histochemistry include alcohol, NAD-linked α-glycerophosphate, glyceraldehyde-3-phosphate (of the Embden–Meyerhof glycolytic pathway), malate, hydroxybutyrate and glucose dehydrogenases. By an ingenious linkage of enzyme systems, Lake (1965) was able to demonstrate fructose-1-phosphate aldolase and fructose-1,6-diphosphate aldolase histochemically and to show the relevance of these methods to a clinical case of fructose intolerance.

$NADH_2$-DIAPHORASE

When a section is given a substrate, AH_2, together with NAD and a suitable hydrogen-acceptor (a tetrazole), the NAD-dependent dehydrogenase oxidizes the substrate and a reduced tetrazole (a coloured formazan) is produced. But this production of the formazan depends on two events: first the substrate and the NAD come to be sited on the active surface of the primary dehydrogenase and hydrogen is transferred from the substrate to the NAD. Thus the substrate becomes oxidized ($AH_2 \rightarrow A$) and the NAD becomes reduced ($NAD \rightarrow NADH_2$). The reduced NAD may remain at the same site, or it may diffuse away. In either event, no formazan will be produced unless the reduced co-enzyme ($NADH_2$) itself becomes a substrate for an enzymic system which oxidizes the $NADH_2$ and transfers the hydrogen to an oxidation-reduction system of an electrode potential compatible with that of the tetrazole (see p. 178). If this occurs, the tetrazole will be reduced, and the section will be coloured by the formazan. (It should be noted that the removal of hydrogen from $NADH_2$ and its donation to a tetrazole or some other dye can be mediated non-enzymically by phenazine methosulphate, by porphyrindines and by porphyrexides, according to Dixon and Webb (1964), p. 367.)

$$AH_2 + NAD \xrightarrow{\text{dehydrogenase}} A + NADH_2$$

$$NADH_2 + \text{tetrazole} \xrightarrow{\text{diaphorase}} NAD + \text{formazan}$$

or $AH_2 + NAD$

↓ dehydrogenase

$NADH_2 + A$

(diaphorase: $NADH_2$ → NAD, cycling back)

+ tetrazole

↓

formazan

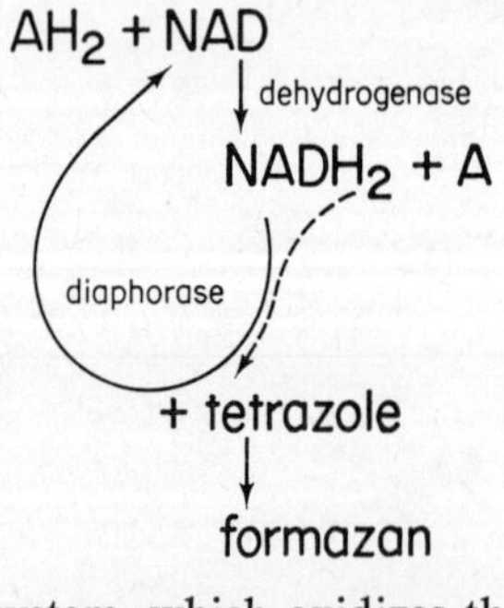

This enzymic system, which oxidizes the $NADH_2$ back to NAD and is also instrumental in reducing the tetrazole, is what is meant by '$NADH_2$-diaphorase'. It is likely that, in completely intact tissue, the dehydrogenases will be close to a diaphorase system; at least this would seem reasonable for the economy of the cell. But in histochemical reactions, this proximity is rarely proven and the histochemist has to test whether the same localization of the formazan would not be produced if the $NADH_2$ were to have been freely diffusable. Equally he may wish to know whether some disturbance of the diaphorase may not account for his obtaining a negative reaction for the primary dehydrogenase. To test this, all he need do is to incubate a serial section in the presence of $NADH_2$ and the tetrazole.

METHOD

Use fresh cryostat sections, preferably serial to those used for the NAD-dependent dehydrogenase, and in the same reaction medium but add 5 mg/ml of reduced nicotinamide-adenine dinucleotide ($NADH_2$) instead of the substrate and the NAD. Incubate at 37°C. Short incubation times will suffice for such a high concentration of the reduced co-enzyme. To simulate the normal incubation, a tenth of this concentration of $NADH_2$ should be sufficient.

NADP-DEPENDENT DEHYDROGENASES

Of the oxidative enzymes which are apparently totally dependent on NADP, those most relevant to histochemists are the NADP-dependent isocitrate dehydrogenase (of the Krebs' cycle, Fig. 25) and the two pentose-shunt dehydrogenases (Fig. 26). The latter, which have been intensely studied, are taken as typical examples from which the histochemistry of NADP-dependent dehydrogenases can be derived.

THE PENTOSE-SHUNT DEHYDROGENASES

GLUCOSE-6-PHOSPHATE AND 6-PHOSPHOGLUCONATE DEHYDROGENASES

BIOCHEMICAL DATA

The normal oxidation of glucose can go along one of two paths. In the best known, and most commonly found pathway (Embden–Meyerhof) the glucose (containing six carbon atoms) is phosphorylated and then split into two halves, each of three carbon atoms (triose phosphate) which are then oxidized (by triose phosphate dehydrogenase, more correctly known as glyceraldehyde-3-phosphate dehydrogenase; see Fig. 25) ultimately to pyruvate. This can then be fed into the Krebs' cycle, or be reduced to lactate (Fig. 25). The alternative pathway is a 'shunt' mechanism in which the phosphorylated glucose (glucose-6-phosphate) is immediately oxidized, ultimately to yield 6-phosphogluconate; this is then itself oxidized and also one carbon atom is removed as carbon dioxide, to give a sugar phosphate in which the sugar contains only five carbon atoms (i.e. a pentose). Because of this, the whole mechanism is often known as the pentose-shunt (also called the hexose-monophosphate shunt, Fig. 26).

The significance of this shunt in cellular metabolism is two-fold. First, it produces pentose sugars which can be used to produce ribose and deoxyribose for the corresponding nucleic acids (or which can be built into more complex sugars); secondly, these two dehydrogenases are unusual in requiring NADP as co-enzyme and so are a major source of $NADPH_2$. This reduced co-enzyme is of considerable importance in a number of essential metabolic reactions. Some examples deserve to be considered.

(*i*) *Folate reductase* is necessary for the reductive conversion of folic acid co-enzymes to their physiologically active tetra-hydro forms; this conversion requires $NADPH_2$. This enzyme has become of great interest recently both because of its significance in the synthesis of deoxyribonucleic acid and

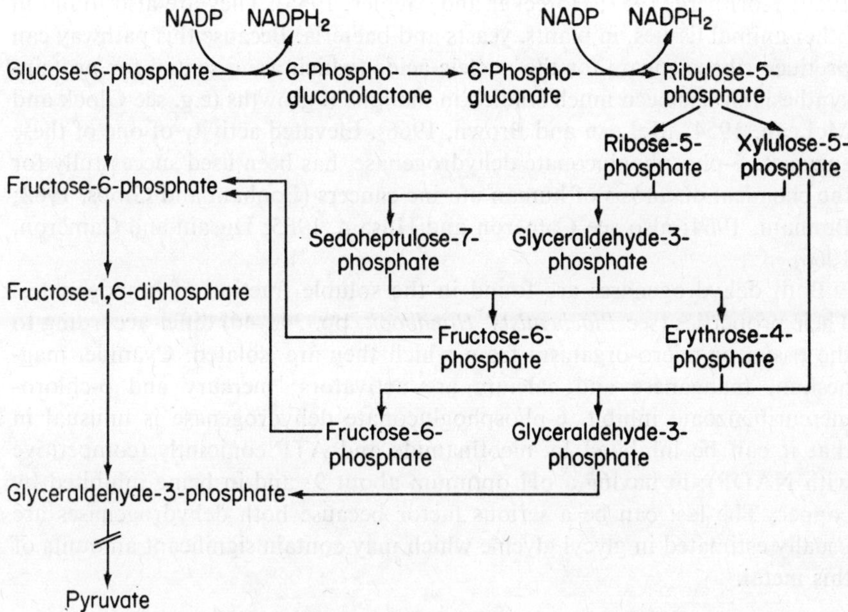

Fig. 26. The pentose-shunt and its various pathways in relation to the Embden–Meyerhof glycolytic pathway (on the left).

because it is inhibited by methotrexate; its activity and inhibition appear to be of considerable significance in leukaemia. A useful histochemical method for it has been suggested by Gunlack *et al.* (1966).

(*ii*) *The synthesis of steroids*, and their interconversion into physiologically important molecules, requires $NADPH_2$ (see Popjak, 1961; Savard, 1961). In view of this, it is not surprising that the adrenal cortex contains high concentrations of pentose-shunt dehydrogenases.

(*iii*) *The synthesis of fatty acids* (see Kornberg, 1961) and of glycogen (Chain, 1959).

(*iv*) Parts of the pentose-shunt play a major role in *photosynthesis* (Krebs and Kornberg, 1957).

(*v*) More recently it has been shown that the *hydroxylation* (and *detoxication*) of many lipid-soluble drugs and of steroid hormones is mediated by $NADPH_2$ and a cytoplasmic cytochrome known as cytochrome P450, which acts as a terminal oxidase (Orrhenius and Ernster, 1967).

It is clear, therefore, that the pentose-shunt dehydrogenases are of considerable significance in cellular metabolism. They occur in high concentration in liver, adrenal, kidney and in the lactating mammary gland; much of their biochemistry has been elucidated in the first and last tissue (Glock and McLean, 1953, 1958); the subject has been well reviewed by Dickens, 1956; Kornberg, 1957; Horecker and Mehler, 1955). They are also found in other animal tissues, in plants, yeasts and bacteria. Because this pathway can produce ribose sugars for the nucleic acids (which are essential for protein-synthesis) it has been much studied in malignant growths (e.g. see Glock and McLean, 1954; McLean and Brown, 1966). Elevated activity of one of these enzymes, 6-phosphogluconate dehydrogenase, has been used successfully for the chemical diagnosis of human uterine cancers (Bonham and Gibbs, 1962; Bonham, 1964; also see Cameron and Husain, 1965; Husain and Cameron, 1966).

Both dehydrogenases are found in the soluble fraction of homogenates. Their properties (see *Biochemists' Handbook*, pp. 343–46) differ according to the tissue or micro-organism from which they are isolated. Cyanide, magnesium, manganese and calcium are activators; mercury and *p*-chloromercuribenzoate inhibit. 6-phosphogluconate dehydrogenase is unusual in that it can be inhibited by nicotinamide and ATP conjointly (competitive with NADP); in having a pH optimum about 9; and in being inhibited by copper. The last can be a serious factor because both dehydrogenases are usually estimated in glycyl glycine which may contain significant amounts of this metal.

HISTOCHEMICAL DATA

It was the histochemical demonstration of exceptionally high activity of 6-phosphogluconate dehydrogenase in cancer (Chayen *et al.*, 1962) which led to the biochemical test for the diagnosis of cervical cancer. It also provoked a histochemical stain for assisting in the cytological screening of human cervical cancer (Cohen and Way, 1966).

These dehydrogenases have been a critical test for histochemistry. Biochemical evidence shows that these enzymes occur in the soluble fraction of homogenates; the histochemical data indicate that they are very weakly bound in the cytoplasm (Altman and Chayen, 1966). Consequently they are readily lost from tissue sections into the incubation medium. If this medium is added as a drop to the section, placed horizontally, the enzymes escape into the drop of liquid, act on the substrate in the medium and generate $NADPH_2$ which then acts as the substrate for the $NADPH_2$–diaphorase system (see later); this system remains bound inside the section and therefore precipitates reduced tetrazolium (formazan) in the section. This is clearly unsatisfactory. Altman (1972) has made a detailed study of the histochemistry of these dehydrogenases and has shown that they can be completely retained inside the sections, in a fully active state, if a suitable concentration of polyvinyl alcohol (PVA) is included in the reaction medium. The concentration of the reactants must be altered to take into account their slower diffusion through the PVA; when this is done, the activity of these enzymes in the

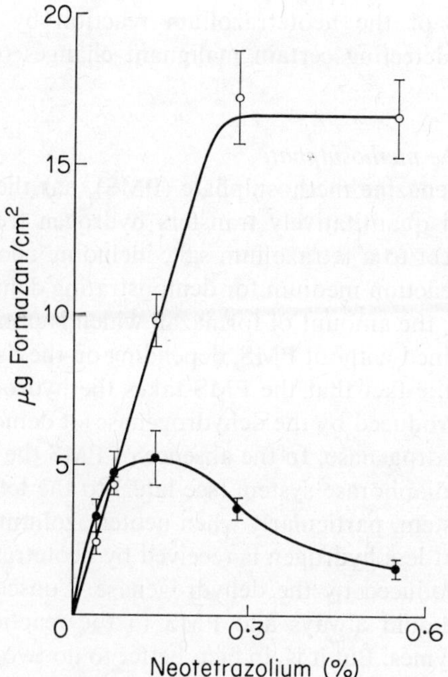

Fig. 27. Activity of glucose-6-phosphate dehydrogenase (rat liver) with various concentrations of neotetrazolium. Filled circles: in the absence of PVA; open circles: in the presence of 20% PVA.

section is considerably greater than in unprotected sections (Fig. 27) and is equivalent to the activity found biochemically. In the protected sections, there is no inhibition by increased concentrations of neotetrazolium. Butcher (1971b) showed that the correct concentration of a collagen polypeptide (40% w/v of collagen polypeptide 5163, available from Sigma) had equally good protective properties. The significance of the use of these inert colloid stabilizers has been discussed in Part II.

The apparently inhibitory effect of oxygen has been studied by Altman (1970). Although the precise mechanism is still not fully understood, oxygen has a marked inhibitory effect on the amount of formazan produced when neotetrazolium is used as the hydrogen-acceptor for these dehydrogenases but it has no effect with tetrazoles of more negative electrode potential (MTT, INT, TNBT or nitro-blue tetrazolium). It seems reasonable to assume that the inhibitory effect of oxygen is due to competition between it and neotetrazolium (and other tetrazoles of more positive electrode potential than nitro-blue tetrazolium). This question has been fully discussed by Altman (1972). In practice, therefore, when testing dehydrogenase activity it is best to saturate the reaction medium with nitrogen. A point of some interest, however, is the fact that many types of malignant cell do not show such striking inhibition of the neotetrazolium reaction by oxygen; this may become useful in detecting certain malignant changes (e.g. Altman *et al.*, 1970).

The use of phenazine methosulphate

This substance, phenazine methosulphate (PMS), has the valuable property that it rapidly and quantitatively transfers hydrogen from $NADPH_2$ (and from $NADH_2$) direct to a tetrazolium salt, including neotetrazolium. When it is added to the reaction medium for demonstrating dehydrogenase enzyme activity in sections, the amount of formazan which is obtained may be up to ten times that obtained without PMS, depending on the tissue. This enhanced staining is due to the fact that the PMS takes the hydrogen directly off the $NADPH_2$ as it is produced by the dehydrogenase; it demonstrates the actual activity of the dehydrogenase. In the absence of PMS the hydrogen is passed via the $NADPH_2$–diaphorase system (see later) to the tetrazolium salt used. The diaphorase system, particularly when neotetrazolium is used, is severely rate-limiting so that less hydrogen is received by neotetrazolium in unit time than is actually produced by the dehydrogenase. Consequently it might be argued that one should always add PMS to the reaction medium for all dehydrogenase enzymes. But it is, in fact, better to do two reactions, one with and the other without PMS because each gives important information. It has been suggested (Chayen *et al.*, 1972; Altman, 1972) that there are two types or two pathways of hydrogen, each having its own particular physiological significance. The hydrogen, from $NADPH_2$, which is passed through the $NADPH_2$–diaphorase system (Type I) may be related to the microsomal respiratory chain (see Gillette *et al.*, 1969). The reaction with PMS shows the total production of $NADPH_2$; this minus Type I gives the residue (Type II) which does not pass down the microsomal respiratory chain and may be that $NADPH_2$ hydrogen which remains at a very negative electrode potential for use, for example, in cellular biosynthetic mechanisms.

METHOD (after Altman, 1972)

Fresh cryostat sections of unfixed tissue must be used.

1. Incubate in the reaction medium in rings in an atmosphere of nitrogen; 20 min at 37°C should be sufficient for most tissues although 10 min may be ample when PMS is added to the reaction medium.

2. Wash in distilled water.
3. Mount in Farrants' medium.

Result. Activity due to pentose-shunt dehydrogenation is shown by the deposition of the formazan (red-purple stain if nitro-blue tetrazolium is used; grains and red colour if neotetrazolium is employed).

Control. Incubate as for the test but omit either the substrate, or the NADP, from the reaction medium.

Reaction medium
Dissolve 5mM of a ditetrazole or 10mM of a monotetrazolium salt (e.g. 0·3 g of neotetrazolium chloride of good purity) in 100 ml of a 0·05M glycyl glycine-sodium hydroxide buffer. At this stage the pH is not critical; it should be about 8·0. If necessary, warm gently to dissolve. Cool and filter. This solution should be kept in the refrigerator and should be stable for some weeks.

When required, pour the required volume of this solution into a beaker which should be of approximately twice the volume of the solution. Slowly sprinkle PVA into it (10 g to 50 ml of the solution) and stir with a glass rod. When most of it is dissolved (which it will do if enough patience is bestowed on the task), leave it to stand for 10 min. Then continue stirring, breaking up any lumps that may have formed. The new grade of PVA, B05/140, dissolves much more easily than did the original M05/140 grade.

When dissolved, add 3mM of NADP (2·5 mg/ml of the disodium salt) and either 5mM of glucose-6-phosphate (1·5 mg/ml of the disodium salt) or 8mM of phosphogluconate (3 mg/ml of the trisodium salt) depending on the enzyme to be tested. At this stage adjust the pH as required: for glucose-6-phosphate dehydrogenase it should be pH 8·0, for 6-phosphogluconate dehydrogenase, it should be pH 7·6. Warm to 37°C and bubble nitrogen through this reaction medium, at 37°C, for 2 min before use.

If it is required to test the total activity of these dehydrogenases, add PMS at concentrations of 0·2 mg/ml for either enzyme. It is advisable to add this last of all because it rapidly decomposes.

$NADPH_2$–DIAPHORASE

In the past, biochemists have recognized the existence of diaphorases which oxidize $NADPH_2$ and $NADH_2$; the diaphorase prepared from pig heart is a powerful lipoic dehydrogenase (*Biochemists' Handbook*, p. 356). Other enzymes which may act as diaphorases are $NADH_2$–cytochrome *c* reductase and $NADPH_2$-cytochrome *c* reductase. In recent years there has been a great surge of interest following the discovery that the microsomes of many, if not all, tissues have a powerful $NADPH_2$-oxidizing system and a respiratory chain (Fig. 28) which passes hydrogen both from $NADPH_2$ and from

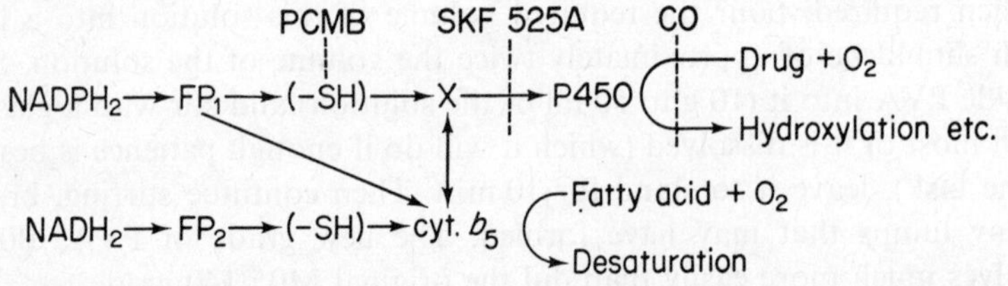

Fig. 28. A simplified scheme showing the generalized microsomal respiratory system, as it is understood at present. Cytochrome *P*-450 is involved in drug detoxication and hydroxylation; it can be inhibited by carbon monoxide (CO). The inhibitor SKF 525A acts before this cytochrome in the respiratory chain, and *para*-chloromercuribenzoate (PCMB) inhibits sulphydryl components of the chain. There appear to be various points (indicated by arrows) at which reducing equivalents can be transferred between the NADPH and the NADH respiratory systems. FP_1 and FP_2 are two flavoproteins.

$NADH_2$ to react with atmospheric oxygen by means of a cytoplasmic cytochrome, cytochrome *P*-450. The special interest of this system is its involvement in the detoxication of many drugs, including the barbiturates, codeine, aminopyrine and chlorpromazine; in the hydroxylation of many steroids and of certain carcinogenic hydrocarbons; in the ω-oxidation of fatty acids; in the mixed function amine oxidase activity which uses $NADPH_2$ for the oxygen-dependent oxidation of certain physiologically important amines such as those used as tranquillizers, as antihistaminics, and of narcotics, topine alkaloids, ephedrine and related compounds. It is apparent that this microsomal respiratory system is of considerable significance in toxicology and pharmacology. The $NADH_2$-microsomal respiratory system, which seems to be coupled with that which uses $NADPH_2$, is involved in the desaturation of fatty acids. A further important effect of the microsomal respiratory system is that it may hydroxylate otherwise inert compounds into necrotizing agents; this is a good example of what Sir Rudolph Peters (1963) called 'lethal synthesis'. The function of the microsomal respiratory system has been reviewed by Gillette *et al.* (1969); its relation to cellular biology and histochemistry is discussed fully by Chayen *et al.* (1972).

For the histochemist, however, the important feature is that he cannot obtain a formazan from the oxidation of glucose-6-phosphate or of 6-phosphogluconate unless there is some system which will oxidize the reduced co-enzyme at the expense of the tetrazolium salt (Fig. 29). This system can be

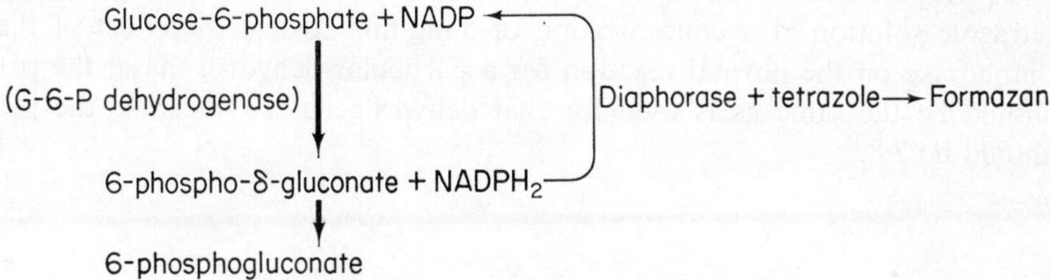

Fig. 29. Formazan is produced only by the mediation of the 'diaphorase' system (in the absence of PMS).

the microsomal respiratory chain (Fig. 28) or some other '$NADPH_2$–diaphorase'; alternatively he may use phenazine methosulphate (PMS) to remove hydrogen directly from the reduced NADP and transmit it to the tetrazole salt which he uses (see above, under 'The Pentose-shunt Dehydrogenases', p. 208). But, in view of the possible importance of the oxidation of $NADPH_2$, he may well wish to test this separately. Admittedly when he does a reaction for glucose-6-phosphate or for 6-phosphogluconate dehydrogenase with neotetrazolium as the final hydrogen-acceptor, he is testing the net result of the dehydrogenase activity modified by the 'diaphorase' activity. He may require to know how active the 'diaphorase' system can be under optimal conditions.

METHOD (Altman, 1972)

Fresh cryostat sections of unfixed tissue must be used.

1. Incubate in the reaction medium in rings in an atmosphere of nitrogen; 20 min at 37°C should be sufficient.
2. Wash in distilled water.
3. Mount in Farrants' medium.

Result. A deposit of the coloured formazan indicates diaphorase activity.

Control. Omit the $NADPH_2$. Or add mercury (10^{-3}M mercuric chloride) which should inhibit $NADPH_2$-diaphorase activity.

Reaction medium

To 0·05M glycylglycine–sodium hydroxide buffer is added either a ditetrazolium salt (5mM) or a monotetrazolium salt (10mM), i.e. 0·3 g of neotetrazolium chloride, of good purity, to 100 ml of the buffer. To 100 ml of this solution is added 20 g of polyvinyl alcohol (see method for glucose-6-phosphate dehydrogenase, p. 209). This stock solution can be stored at +4°C. [*Note*: unstable tetrazoles cannot be incorporated in the stock solution and have to be added just before use; neotetrazolium and nitro-blue tetrazolium can be made up in the stock solution.]

To prepare the reaction medium, $NADPH_2$ is added to the PVA-buffer-tetrazole solution at a concentration of 5 mg/ml. To test the effect of the diaphorase on the normal reaction for a particular dehydrogenase, the pH should be the same as is used for that dehydrogenase. Otherwise the pH should be 7·8.

TRANSHYDROGENASES

Another way by which the hydrogen from $NADPH_2$ can be transferred to a tetrazole is by transhydrogenation mechanisms. These have been extensively reviewed by Villee (1962). They are of considerable physiological interest in that they involve the direct intervention of steroid hormones. Other mechanisms by which hydrogen, liberated from substrate in the cytoplasm, can be transferred to the mitochondrial hydrogen-transport system, have been discussed by Lehninger (1965). The work of Butcher and Chayen (1966b) indicated that histochemical methods may prove more valuable than more conventional biochemical procedures for studying such phenomena; this question has been discussed in some detail by Chayen and Bitensky (1968).

REFERENCES

ADAMS, C. W. M. 1961. A perchloric acid-naphthoquinone method for the histochemical localization of cholesterol. *Nature*, **192**, 331–32.

ADAMS, C. W. M. Ed. 1965. *Neurohistochemistry*. Elsevier, Amsterdam.

ADAMS SMITH, W. N., and STOWARD, P. J. 1967. A fluorescence histochemical study of ketosteroids in gonadal and adrenal tissue of the rat. *J. Roy. micr. Soc.*, **87**, 47–52.

ADLAM, G. H. J., and PRICE, L. S. 1938. *Inorganic Chemistry*. John Murray, London.

ALDRIDGE, W. N. 1956. Organophosphorus compounds and esterases. *Ann. Rep. Chem. Soc.*, **53**, 294–305.

ALDRIDGE, W. N. 1961. Esterases. In *Biochemists' Handbook*. Long, C., Ed. pp. 273–77. Spon, London.

ALFERT, M., and GESCHWIND, I. I. 1953. A selective staining method for the basic proteins of cell nuclei. *Proc. Nat. Acad. Sci.*, **39**, 991–99.

ALI, S. Y. 1964. The degradation of cartilage matrix by an intracellular protease. *Biochem. J.*, **93**, 611–18.

ALTMAN, F. P. 1969a. The quantitative elution of nitro-blue formazan from tissue sections. *Histochemie*, **17**, 319–26.

ALTMAN, F. P. 1969b. The use of eight different tetrazolium salts for a quantitative study of pentose-shunt dehydrogenation. *Histochemie*, **19**, 363–74.

ALTMAN, F. P. 1970. On the oxygen-sensitivity of various tetrazolium salts. *Histochemie*, **22**, 256–61.

ALTMAN, F. P. 1971. The use of a recording microdensitometer for the quantitative measurement of enzyme activities inside tissue sections. *Histochemie*, **27**, 125–36.

ALTMAN, F. P. 1972. Quantitative dehydrogenase histochemistry with special reference to pentose-shunt dehydrogenases. *Progr. in Histochem. Cytochem.* Fischer, Stuttgart.

ALTMAN, F. P., BITENSKY, L., BUTCHER, R. G., and CHAYEN, J. 1970. Integrated cellular chemistry applied to malignant cells. In *Cytology Automation*. Evans, D. M. D., Ed. Livingstone, Edinburgh.

ALTMANN, F. P., and CHAYEN, J. 1965. Retention of nitrogenous material in unfixed sections during incubation for histochemical demonstration of enzymes. *Nature*, **207**, 1205–6.

ALTMANN, F. P., and CHAYEN, J. 1966. The significance of a functioning hydrogen-transport system for the retention of 'soluble' dehydrogenases in unfixed sections. *J. Roy. micr. Soc.*, **85**, 175–80.

ANDERSEN, H., MØLLGÅRD, K., and BÜLOW, F. A. VON. 1970. On the specificity of staining by Alcian blue in the study of human foetal adenohypophysis. *Histochemie*, **22**, 362–75.

ANDERSON, P. J. 1967. Purification and quantitation of gluteraldehyde and its effect on several enzyme activities in skeletal muscle. *J. Histochem. Cytochem.*, **15**, 652–61.

APPLETON, T. C. 1964. Autoradiography of soluble labelled compounds. *J. Roy. micr. Soc.*, **83**, 277–82.

ARBORGH, B., ERICSSON, J. L. E., and HELMINEN, H. 1971. Inhibition of renal acid phosphatase and aryl sulfatase activity by glutaraldehyde fixation. *J. Histochem. Cytochem.*, **19**, 449–51.

ARMSTRONG, J. A. 1956. Histochemical differentiation of nucleic acids by means of induced fluorescence. *Exptl Cell Res.*, **11**, 640–43.

AUSTIN, C. R., and AMOROSO, E. C. 1959. The mammalian egg. *Endeavour*, **18**, 130.

BAILLIE, A. H., FERGUSON, M. M., and HART, D. MCK. 1966. *Developments in Steroid Histochemistry*. Academic Press, London.

REFERENCES

BAKER, J. R. 1946. The histochemical recognition of lipine. *Quart. J. micr. Sci.*, **87**, 441–70.

BAKER, J. R. 1947. The histochemical recognition of certain guanidine derivatives. *Quart. J. micr. Sci.*, **88**, 115–21.

BAKER, J. R. 1956. The histochemical recognition of certain phenols, especially tyrosine. *Quart. J. micr. Sci.*, **97**, 161–64.

BAKER, J. R. 1958. *Principles of Biological Microtechnique*. Methuen, London.

BALDWIN, E. 1959. *Dynamic Aspects of Biochemistry*. 3rd edn. Univ. Press, Cambridge.

BALL, J., and JACKSON, D. S. 1954. Histological, chromatographic and spectrophotometric studies of toluidine blue. *Stain Technol.*, **28**, 33–40.

BARKA, T. 1960. A simple azo-dye method for histochemical demonstration of acid phosphatase. *Nature*, **187**, 248–49.

BARKA, T., and ANDERSON, P. J. 1963. *Histochemistry*. Harper and Row, New York.

BARRNETT, R. J., and SELIGMAN, A. M. 1951. Histochemical demonstration of esterases by production of indigo. *Science*, **114**, 579.

BARRNETT, R. J., and SELIGMAN, A. M. 1952. Demonstration of protein-bound sulfhydryl and disulfide groups by two new histochemical methods. *J. Nat. Canc. Inst.*, **13**, 215–16.

BARRNETT, R. J., and SELIGMAN, A. M. 1954. Histochemical demonstration of sulfhydryl and disulfide groups of protein. *J. Nat. Canc. Inst.*, **14**, 769–92.

BARTER, R., DANIELLI, J. F., and DAVIES, H. G. 1955. A quantitative cytochemical method for estimating alkaline phosphatase activity. *Proc. Roy. Soc. Lond., B*, **144**, 412–26.

BAUDHUIN, P., BEAUFAY, H., RAHMAN-LI, Y., SELLINGER, O. Z., WATTIAUX, R., JACQUES, P., and DUVE, C. DE. 1964. Tissue fractionation studies. 17. Intracellular distribution of monoamine oxidase, aspartate aminotransferase, alanine aminotransferase, D-amino acid oxidase and catalase in rat-liver tissue. *Biochem. J.*, **92**, 179–84.

B.D.H. booklet 1962. *The Colorimetric Determination of Oxidation-Reduction Balance*. British Drug Houses Ltd., Poole.

BEAUFAY, H., and DUVE, C. DE. 1954. Le système hexose-phosphatasique. VI. Essais de démembrement des microsomes portents de glucose-6-phosphatase. *Bull. Soc. Chim. biol. Paris*, **36**, 1551–68.

BEAUFAY, H., JACQUES, P., BAUDHUIN, P., SELLINGER, O. Z., BERTHET, J., and DUVE, C. DE. 1964. Tissue fractionation studies. 18. Resolution of mitochondrial fractions from rat liver into three distinct populations of cytoplasmic particles by means of density equilibration in various gradients. *Biochem. J.*, **92**, 184–205.

BEHAL, F. J., LITTLE, G. H., and KLEIN, R. A. 1969. Arylamidases of human liver. *Biochim. Biophys. Acta*, **178**, 118–27.

BELL, L. G. E. 1956. Freeze-drying. In *Physical Techniques in Biological Research*. Oster, G., and Pollister, A. W., Eds. Vol. III, Cells and Tissues. Academic Press, New York.

BENDALL, D. S., and DUVE, C. DE. 1960. Tissue fractionation studies. 14. The activation of latent dehydrogenases in mitochondria from rat liver. *Biochem. J.*, **74**, 444–50.

BENNETT, H. S., and WATTS, R. M. 1958. The cytochemical demonstration and measurement of sulfhydryl groups by azo-aryl mercaptide coupling, with special reference to Mercury Orange. In *General Cytochemical Methods*. Danielli, J. F. Ed. Academic Press, New York.

BENSLEY, R. R., and GERSH, I. 1933. Studies on cell structure by the freezing–drying method. II The nature of mitochondria in the hepatic cell of amblyostoma. *Anat. Rec.*, **57**, 217–33.

BERENBAUM, M. C. 1958. The histochemistry of bound lipids. *Quart. J. micr. Sci.*, **99**, 231–42.

BERENBOM, M., YOKOYAMA, H. O., and STOWELL, R. E. 1952. Chemical and enzymatic changes in liver following freezing-drying and acetone fixation. *Proc. Soc. Exptl Biol. Med.*, **81**, 125–28.

BERG, N. O. 1951. A histological study of masked lipids. *Acta path. Microbiol. Scand., Suppl.* **40**.

BERGMANN, M. 1942. A classification of proteolytic enzymes. *Adv. Enzymol.*, **2**, 49–68.

REFERENCES

Bergmann, M., and Fruton, J. S. 1941. The specificity of proteinases. *Adv. Enzymol.*, **1,** 63–98.

Bernfeld, P., Bernfeld, H. C., Nisselbaum, J. S., and Fishman, W. H. 1954. Dissociation and activation of β-glucuronidase. *J. Am. Chem. Soc.*, **76,** 4872–77.

Biochemists' Handbook. 1961. Long, C., Ed. Spon, London.

Bitensky, L. 1962. The demonstration of lysosomes by the controlled temperature freezing-sectioning method. *Quart. J. micros. Sci.*, **103,** 205–9.

Bitensky, L. 1963a. The reversible activation of lysosomes in normal cells and the effects of pathological conditions. In *Ciba Symp. on 'Lysosomes'*. de Reuck, A. V. S., and Cameron, M. P., Eds. Churchill, London.

Bitensky, L. 1963b. Cytotoxic action of antibodies. *Brit. Med. Bull.*, **19,** 241–44.

Bitensky, L. 1963c. Modifications to the Gomori acid phosphatase technique for controlled-temperature frozen sections. *Quart. J. micros. Sci.*, **104,** 193–96.

Bitensky, L. 1967a. Histochemistry of liver disease. In *The Liver*. Read, A. E., Ed. Colston Res. Soc. Symp. Butterworth, London.

Bitensky, L. 1967b. Histochemistry in experimental immunology. In *Handbook of Experimental Immunology*. Weir, D. M., Ed. Blackwell, Oxford.

Bitensky, L., Butcher, R. G., and Chayen, J. 1972. Quantitative cytochemistry in the study of lysosomal function. In *Lysosomes in Biology and Pathology*. Dingle, J. T., Ed. Vol. III. North Holland, Amsterdam.

Bitensky, L., Chayen, J., Cunningham, G. J., and Fine, J. 1963. Behaviour of lysosomes in haemorrhagic shock. *Nature*, **199,** 493–94.

Bitensky, L., and Cohen, S. 1965. The variation of endometrial acid phosphatase activity with the menstrual cycle. *J. Obstet. Gynaec. Brit. Comm.*, **72,** 769–74.

Bitensky, L., Ellis, R., Silcox, A. A., and Chayen, J. 1962. Histochemical studies on carbohydrate material in liver. *Ann. Histochim.*, **1,** 9–14.

Bitensky, L., and Gahan, P. B. 1962. The reversible activation of lysosomes in normal cells. *Biochem. J.*, **84,** 13P

Björklund, A., Ehinger, B., and Falck, B. 1968. A method for differentiating dopamine from noradrenaline in tissue sections by microspectrofluorometry. *J. Histochem. Cytochem.*, **16,** 263–70.

Björklund, A., and Stenevi, U. 1970. Acid catalysis of the formaldehyde condensation reaction for sensitive histochemical demonstration of tryptamines and 3-methoxylated phenylethylamines. I. Model experiments. *J. Histochem. Cytochem.*, **18,** 794–882.

Blaschko, H. 1952. Amine oxidase and amine metabolism. *Pharmacol. Rev.*, **4,** 415–58.

Blaschko, H. 1961. Amine oxidases. In *Biochemists' Handbook*. Long, C., Ed. pp. 373–75. Spon, London.

Bonham, D. G. 1964. A new test for the diagnosis of gynaecological cancer, 6-phosphogluconate dehydrogenase activity in vaginal fluid. *Triangle*, **6,** 157–62.

Bonham, D. G., and Gibbs, D. F. 1962. A new enzyme test for gynaecological cancer: 6-phosphogluconate dehydrogenase activity in vaginal fluid. *Brit. Med. J.*, ii, 823–24.

Bonner, J. 1950. *Plant Biochemistry*. Academic Press, New York.

Braimbridge, M. V., Darracott, S., Chayen, J., Bitensky, L., and Poulter, L. W. 1967. Possibility of a new infective aetiological agent in congestive cardiomyopathy. *Lancet*, i, 171–76.

Brachet, J. 1954. The use of basic dyes and ribonuclease for the cytochemical detection of ribonucleic acid. *Quart. J. micr. Sci.*, **94,** 1–10.

Brown, D. M., and Todd, A. R. 1955. Evidence on the nature of the chemical bonds in nucleic acids. In *The Nucleic Acids*. Chargaff, E., and Davidson, J. N., Eds. Vol. I, pp. 409–46. Academic Press, New York.

Bruyn, P. P. H. de, Farr, R. S., Banks, H., and Morthland, F. W. 1953. *In vivo* and *in vitro* affinity of diaminoacridines for nucleoproteins. *Exptl Cell Res.*, **4,** 174.

Burdon, K. L. 1947. *Textbook of Microbiology*. 3rd edn. Macmillan, New York.

Burstone, M. S. 1959. New histochemical techniques for the demonstration of tissue oxidase (cytochrome oxidase). *J. Histochem. Cytochem.*, **7,** 112–22.

REFERENCES

Burstone, M. S. 1960. Histochemical demonstration of cytochrome oxidase with new amine reagents. *J. Histochem. Cytochem.*, **8**, 63–70.

Burstone, M. S. 1962. *Enzyme Histochemistry and its Application in the Study of Neoplasms.* Academic Press, New York.

Burton, K. 1961. Free energy data and oxidation-reduction potentials. *Biochemists' Handbook.* Long, C., Ed. pp. 90 95. Spon, London.

Butcher, R. G. 1968. The estimation of the nucleic acids of tissue sections and its use as a unit of comparison for quantitative histochemistry. *Histochemie*, **13**, 263–75.

Butcher, R. G. 1970. Studies on succinate oxidation. I. The use of intact tissue sections. *Exptl Cell Res.*, **60**, 54–60.

Butcher, R. G. 1971a. The chemical determination of section thickness. *Histochemie*, **28**, 131–36.

Butcher, R. G. 1971b. Tissue stabilization during histochemical reactions: the use of collagen polypeptides. *Histochemie*, **28**, 231–35.

Butcher, R. G., and Chayen, J. 1966a. Quantitative studies on the alkaline phosphatase reaction. *J. Roy. micr. Soc.*, **85**, 111–17.

Butcher, R. G., and Chayen, J. 1966b. Dehydrogenase interactions in integrated tissue biochemistry. *Biochem. J.*, **100**, 47P

Butcher, R. G., Chayen, J., and Labrum, A. H. 1965. Oxidation of L-ascorbic acid by cells of carcinoma of the human cervix. *Nature*, **207**, 992–93.

Butcher, R. G., Diengdoh, J. V., and Chayen, J. 1964. A study of the histochemical demonstration of cytochrome oxidase. *Quart. J. micros. Sci.*, **105**, 497–502.

Cahn, R., Kaplan, N. O., Levine, L., and Zwilling, E. 1962. Nature and development of lactic dehydrogenases. *Science*, **136**, 962–69.

Cain, A. J. 1950. The histochemistry of lipoids in animals. *Biol. Rev.*, **25**, 73–112.

Cameron, C. B., and Husain, O. A. N. 1965. 6-Phosphogluconate dehydrogenase activity in vaginal fluid: limitations as a screening test for genital cancer. *Brit. Med. J.*, i, 1529–30.

Carver, M. J., Brown, F. C., and Thomas, L. E. 1953. An arginine histochemical method using Sakaguchi's new reagent. *Stain Technol.*, **28**, 89–91.

Casselman, W. G. B. 1959. *Histochemical Technique.* Methuen, London.

Chain, E. B. 1959. Recent studies on carbohydrate metabolism. *Brit. Med. J.*, ii, 707–19.

Charnock, J. S., and Opit, L. J. 1968. Membrane metabolism and ion transport. In *The Biological Basis of Medicine.* Bittar, E. E. and N., Eds. Vol. I, 69–104. Academic Press, New York.

Chayen, J. 1952. The methyl green-pyronin method. *Exptl Cell Res.*, **3**, 652–55.

Chayen, J. 1953. Ascorbic acid and its intracellular localization, with special reference to plants. *Int. Rev. Cytol.* **2**, 78–131.

Chayen, J. 1968a. Histochemistry of phospholipids and its significance in the interpretation of the structure of cells In *The Interpretation of Cell Structure.* Ross, K. F., and McGee-Russell, S., Eds. Arnold, London.

Chayen, J. 1968b. Quantitative histochemistry: cell structure revealed through cellular function. In *The Interpretation of Cell Structure*, Ross, K. F., and McGee-Russell, S., Eds. Arnold, London.

Chayen, J., Altmann, F. P., Bitensky, L., Braimbridge, M. V., and Kadas, T. 1966a. A study of the changes in hydrogen transport in an isolated rat heart preparation. *J. Roy. micr. Soc.*, **86**, 151–58.

Chayen, J., Altman, F. P., and Butcher, R. G. 1972. The production and possible utilization of reducing equivalents outside the mitochondria and the effects of certain drugs. In *Fundamentals of Cell Pharmacology.* Dikstein, S., Ed. Thomas, Illinois.

Chayen, J., and Bitensky, L. 1968. The multiphase chemistry of cell injury. In *The Biological Basis of Medicine.* Bittar, E. E., Ed. Vol. I. Academic Press, London.

Chayen, J., and Bitensky, L. 1971. Lysosomal enzymes and inflammation with particular reference to rheumatoid diseases. *Ann. Rheum. Dis.*, **30**, 522–36.

REFERENCES

CHAYEN, J., BITENSKY, L., AVES, E. K., JONES, G. R. N., SILCOX, A. A., and CUNNINGHAM, G. J. 1962. Histochemical demonstration of 6-phosphogluconate dehydrogenase in proliferating and malignant cells. *Nature*, **195,** 714–15.

CHAYEN, J., BITENSKY, L., BUTCHER, R. G., and CASHMAN, B. 1971a. Evidence for altered lysosomal membranes in synovial lining cells from human rheumatoid joints. *Beitr. Path.*, **142,** 137–49.

CHAYEN, J., BITENSKY, L., BUTCHER, R. G., POULTER, L. W., and UBHI, G. S. 1970. Methods for the direct measurement of anti-inflammatory action on human tissue maintained *in vitro*. *Br. J. Derm.*, **82,** Suppl. 6, 62–81.

CHAYEN, J., BITENSKY, L., and POULTER, L. W. 1966. Fluorescence histochemistry of steroids and carotenoids. *Proc. Roy. micr. Soc.*, **1,** 212.

CHAYEN, J., BITENSKY, L., and UBHI, G. S. 1971b. Quantitative cytochemical studies on psoriasis and the effect of certain drugs. *Clin. Trials J.*, **8,** 35–44.

CHAYEN, J., BITENSKY, L., and WELLS, P. J. 1966b. Mitochondrial enzyme latency and its significance in histochemistry and biochemistry. *J. Roy. micr. Soc.*, **86,** 69–74.

CHAYEN, J., CUNNINGHAM, G. J., GAHAN, P. B., and SILCOX, A. A. 1960. Newer methods in cytology. *Bull. Res. Israel*, **8D,** 273–84.

CHAYEN, J., DARRACOTT, S., and KIRKBY, W. W. 1966c. A re-interpretation of the role of the mast cell. *Nature*, **209,** 887–88.

CHAYEN, J., and DENBY, E. F. 1968. *Biophysical Technique as Applied to Cell Biology.* Methuen, London.

CHAYEN, J., and GAHAN, P. B. 1959. An improved and rapid embedding method. *Quart. J. micr. Sci.*, **100,** 275–77.

CHAYEN, J., GAHAN, P. B., and LA COUR, L. F. 1959. The masked lipids of nuclei. *Quart. J. micr. Sci.*, **100,** 325–37.

CHAYEN, J., JONES, G. R. N., BITENSKY, L., and CUNNINGHAM, G. J. 1961. Histological variation as a source of biochemical error. *Biochem. J.*, **79,** 34P.

CHAYEN, R., and ROBERTS, E. R. 1955. Some observations on the metachromatic reaction. *Sci. J. Roy. Coll. Sci.*, **25,** 50–56.

CHEETHAM, R. D., MORRÉ, D. J., and YUNGHANS, W. N. 1970. Isolation of a Golgi apparatus-rich fraction from rat liver. II. Enzymatic characterization and comparison with other cell fractions. *J. Cell Biol.*, **44,** 492–500.

CHÈVREMONT, M., and FIRKET, H. 1951a. Action du béryllium sur la croissance et la mitose en culture de tissus. *Compt. rend. l'Ass. Anat.*, **38,** 1–4.

CHÈVREMONT, M., and FIRKET, H. 1951b. Etude histochimique de l'action du béryllium sur la mitose en culture de tissus (phosphatase alcaline et acides nucléiques). *Compt. rend. Soc. Biol.*, **145,** 938.

CHÈVREMONT, M., and FIRKET, H. 1953. Alkaline phosphatase of the nucleus. *Int. Rev. Cytol.*, **2,** 261–88.

CHÈVREMONT, M., and FRÉDERICK, J. 1943. Une nouvelle méthode histochimique de mise en evidence des substances à fraction sulfhydrile. *Arch. Biol.*, **54,** 589.

CHIQUONE, A. D. 1953. The distribution of glucose-6-phosphatase in the liver and kidney of the mouse. *J. Histochem. Cytochem.*, **1,** 429–35.

CHIQUOINE, A. D. 1955. Further studies on the histochemistry of glucose-6-phosphatase. *J. Histochem. Cytochem.*, **3,** 471–78.

COHEN, S., BITENSKY, L., and CHAYEN, J. 1965. The study of monoamine oxidase activity by histochemical procedures. *Biochem. Pharmacol.*, **14,** 223–26.

COHEN, S., and WAY, S. 1966. Histochemical demonstration of pentose-shunt activity in smears from the uterine cervix. *Brit. Med. J.*, i, 88–89.

COONS, A. H. 1952. The cellular fate of injected antigens. *Symp. Soc. exp. Biol.*, **7,** 166–72.

COONS, A. H. 1956. Histochemistry with labeled antibody. *Int. Rev. Cytol.*, **5,** 1–23.

COONS, A. H. 1958. Fluorescent antibody methods. In *General Cytochemical Methods.* Danielli, J. F., Ed. Academic Press, New York.

COONS, A. H., CREECH, H. J., JONES, R. N., and BERLINER, E. 1942. The determination of pneumococcal antigen in tissues by the use of fluorescent antibody. *J. Immunol.* **45,** 159–70.

REFERENCES

Corrodi, H., and Jonsson, G. 1967. The formaldehyde fluorescence method for the histochemical demonstration of biogenic monoamines. *J. Histochem Cytochem.*, **15,** 65–78.

Cotson, S., and Holt, S. J. 1958. Studies in enzyme cytochemistry. IV. Kinetics of aerial oxidation of indoxyl and some of its halogen derivatives. *Proc. Roy. Soc., B,* **148,** 506–19.

Cunningham, G. J., Bitensky, L., Chayen, J., and Silcox, A. A. 1962. The preservation of cytological and histochemical detail by a controlled temperature freezing and sectioning technique. *Ann. Histochim.*, **7,** 433–35.

Dalcq, A. M. 1962. Localisations et evolution des phosphatases aux premier stades du développement. *Bull. l'Acad. Roy. Med.*, **7,** 573–610.

Dalcq, A. M. 1963. The relation to lysosomes of the *in vivo* metachromatic granules. In *Ciba Symp. on 'Lysosomes'*. de Reuck, A. V. S., and Cameron, M. P., Eds. Churchill, London.

Dalcq, A. M., and Pasteels, J. J. 1963. La localisation d'enzymes de déphosphorylation dans les œufs des quelques Invertébrés. *Developmental Biology,* **7,** 457–87.

Danielli, J. F. 1953. *Cytochemistry: A Critical Approach.* Wiley, New York.

Daoust, R. 1965. Histochemical localization of enzyme activities by substrate film methods; ribonucleases, deoxyribonucleases, protease, amylase and hyaluronidase. *Int. Rev. Cytol.*, **18,** 191–221.

Darlington, C. D., and La Cour, L. F. 1947. *The Handling of Chromosomes.* 2nd edn. Allen and Unwin, London.

Davidson, J. N. 1960. *The Biochemistry of the Nucleic Acids.* 4th edn. Methuen, London.

Davidson, J. N., and Waymouth, C. 1944. The histochemical demonstration of ribonucleic acid in mammalian liver. *Proc. Roy. Soc., Edinburgh,* **62,** 96–98.

Davies, H. G. 1954. The action of fixatives on the ultra-violet-absorbing components of chick fibroblasts. *Quart. J. micr. Sci.*, **95,** 433.

Defendi, V., and Pearson, B. 1955. Quantitative estimation of succinic dehydrogenase activity in a single microscopic tissue section. *J. Histochem. Cytochem.*, **3,** 61–69.

Desai, I. D., and Tappel, A. L. 1963. Damage to proteins by peroxidized lipids. *J. Lipid Res.*, **4,** 204–7.

Desmet, V. J. 1962. The hazard of acid differentiation in Gomori's method for acid phosphatase. *Stain Technol.*, **37,** 373–76.

Dickens, F. 1956. The hexosemonophosphate oxidative pathway of yeast and animal tissues. *3rd Int. Congr. Biochem., Brussels,* 1955. Liébecq, C., Ed. Academic Press, New York.

Diengdoh, J. V. 1964. The demonstration of lysosomes in mouse skin. *Quart. J. micr. Sci.*, **105,** 73–78.

Diengdoh, J. V. 1966. Application of the acridine orange method in the diagnosis of malignancy. *Proc. Roy. micr. Soc.*, **1,** 213.

Diengdoh, J. V. 1966a. The histochemistry of normal skin; its variation under several conditions including those leading up to carcinogenesis. Ph.D. thesis. University of London.

Diengdoh, J. V. 1966b. Histochemical studies on monoamine oxidase in skin. *J. Roy. micr. Soc.*, **85,** 103–9.

Dixon, M., and Webb, E. C. 1964. *Enzymes.* 2nd edn. Longmans, London.

Doyle, W. L. 1948. Effects of dehydrating agents on phosphatases in the lymphatic nodules of the rabbit appendix. *Proc. Soc. Exptl Biol. Med.*, **69,** 43–44.

Dumonde, D. C., Bitensky, L., Cunningham, G. J., and Chayen, J. 1965. The effects of antibodies on cells. 1. Biochemical and histochemical effects of antibodies and complement on ascites tumour cells. *Immunology,* **8,** 25–36.

Duve, C. de 1957. The enzymic heterogeneity of cell fractions isolated by differential centrifuging. *Symp. Soc. Exp. Biol.*, **10,** 50–61.

Duve, C. de 1959. Lysosomes, a new group of cytoplasmic particles. In *Subcellular Particles.* Hayashi, T., Ed. Ronald, New York.

REFERENCES

Duve, C. de, Pressman, B. C., Gianetto, R., Wattiaux, R., and Appelmans, F. 1955. Tissue fractionation studies. VI. Intracellular distribution patterns of enzymes in rat liver tissue. *Biochem. J.*, **60**, 604–17.

Dyer, J. R. 1956. The use of periodate oxidations in biochemical analysis. In *Methods of Biochemical Analysis*, Vol. III. Glick, D., Ed. Interscience, New York.

Ebel, J. P. 1952. Recherches sur les polyphosphates contenus dans diverses cellules vivantes. IV. Localisation cytologique et rôle physiologique des polyphosphates dans la cellule vivante. *Bull. Soc. Chim. Biol.*, **34**, 498–505.

Ellis, S., and Perry, M. 1966. Pituitary arylamidases and peptidases. *J. Biol. Chem.*, **241**, 3479.

Emmel, V. M. 1946. The intracellular distribution of alkaline phosphatase activity following various methods of histologic fixation. *Anat. Rec.*, **95**, 159.

Emmelot, P., and Bos, C. J., 1965. Differential effect of neuraminidase on the magnesium ATP-ase, sodium–potassium–magnesium ATP-ase and 5-nucleotidase of isolated plasma membranes. *Biochim. biophys. Acta*, **99**, 578–80.

Eränkö, O. 1967. The practical histochemical demonstration of catecholamines by formaldehyde-induced fluorescence. *J. Roy. micr. Soc.*, **87**, 259–76.

Eränkö, O., and Härkönen, M. 1965. Monoamine-containing small cells in the superior cervical ganglion of the rat and an organ composed of them. *Acta Physiol. Scand.*, **63**, 511.

Ernst, S. A., and Philpott, C. W. 1970. Preservation of Na–K-activated and Mg-activated adenosine triphosphatase activities of avian salt gland and teleost gill with formaldehyde as fixative. *J. Histochem. Cytochem.*, **18**, 251–63.

Fahimi, H. D., and Drochmans, P. 1968. Purification of glutaraldehyde: its significance for preservation of acid phosphatase activity. *J. Histochem. Cytochem.*, **16**, 199–204.

Farber, E., and Bueding, E. 1956. Histochemical localization of specific oxidative enzymes. V. The dissociation of succinic dehydrogenase from carriers by lipase and the specific histochemical localization of the dehydrogenase with phenazine methosulfate and tetrazolium salts. *J. Histochem. Cytochem.*, **4**, 357–62.

Farber, E., and Morrison, M. 1964. The localization of oxidative enzymes in mitochondria by histochemical staining procedures. *Proc. 2nd Int. Congr. Histochem. Cytochem.* Springer, Berlin.

Farber, E., Sternberg, W. H., and Dunlap, C. E. 1956a. Histochemical localization of specific oxidative enzymes. 1. Tetrazolium stains for diphosphopyridine nucleotide diaphorase and triphosphopyridine nucleotide diaphorase. *J. Histochem. Cytochem.*, **4**, 254–66.

Farber, E., Sternberg, W. H., and Dunlap, C. E. 1956b. Histochemical localization of specific oxidative enzymes. III. Evaluation studies of tetrazolium staining methods for diphosphopyridine nucleotide diaphorase, triphosphopyridine nucleotide diaphorase and the succindehydrogenase system. *J. Histochem. Cytochem.*, **4**, 284–94.

Fiala, S., and Fiala, A. E. 1959. On the correlation between metabolic and structural changes during carcinogenesis in rat liver. *Brit. J. Canc.*, **13**, 136–51.

Fialkow, P. J., and Fishman, W. H. 1961. Studies on a liver activator of β-glucuronidase. *J. Biol. Chem.*, **236**, 2169–71.

Fieser, L. F., and Fieser, M. 1953. *Organic Chemistry*. 2nd edn. Harrap, London.

Fisher, E. R., and Lillie, R. D. 1954. The effect of methylation on basophilia. *J. Histochem. Cytochem.*, **2**, 81.

Fishman, W. H. 1951. The relationship of the enzyme beta-glucuronidase to cancer of the cervix uteri. *Bull. New Eng. Med. Centr.*, **13**, 12–19.

Fishman, W. H. 1961. Renal β-glucuronidase response to steroids of the androgen series. In *Mechanism of Action of Steroid Hormones*. Pergamon, Oxford.

Fishman, W. H. 1963. Recent studies on β-glucuronidase. *Il. Farmaco*, **18**, 397–407.

Fishman, W. H., and Anlyan, A. J. 1947. β-glucuronidase activity in human tissues. Some correlations with processes of malignant growth and with the physiology of reproduction. *Canc. Res.*, **7**, 808–17.

REFERENCES

FISHMAN, W. H., and ANLYAN, A. J. 1950. Beta-glucuronidase activity in human tissues; some correlations with processes of malignant growth and with the physiology of reproduction. *Acta Un. Int. Contr. Canc.*, **6,** 1034–41.

FISHMAN, W. H., and BAKER, J. R. 1956. Cellular localization of β-glucuronidase in rat tissues. *J. Histochem. Cytochem.*, **4,** 570–87.

FISHMAN, W. H., and BIGELOW, R. 1950. A comparative study of the morphology and glucuronidase activity in 44 gastrointestinal neoplasms. *J. Nat. Canc. Inst.*, **10,** 1115–22.

FISHMAN, W. H., and GOLDMAN, S. S. 1965. A postcoupling technique for β-glucuronidase employing the substrate, naphthol AS-BI-β-O-glucosiduronic acid. *J. Histochem. Cytochem.*, **13,** 441–47.

FISHMAN, W. H., GOLDMAN, S. S., and GREEN, S. 1964. Several biochemical criteria for evaluating β-glucuronidase localization. *J. Histochem. Cytochem.*, **12,** 239–51.

FISHMAN, W. H., and LIPKIND, J. B. 1958. Comparative ability of some steroids and their esters to enhance the renal β-glucuronidase activity of mice. *J. Biol. Chem.*, **232,** 729–36.

FITZGERALD, P. J. 1959. Autoradiography in cytology. In *Analytical Cytology*. Mellors, R. C., Ed. 2nd edn. McGraw-Hill, New York.

FLEISCHER, B., FLEISCHER, S., and OZAWA, H. 1969. Isolation and characterization of Golgi membranes from bovine liver. *J. Cell Biol.*, **43,** 59–79.

FOLIN, O., and MARENZI, A. D. 1929. Tyrosine and typtophane determinations in one-tenth gram of protein. *J. Biol. Chem.*, **83,** 89–102.

FREDRICSSON, B. 1958. Preservation of activities of alkaline phosphatase and naphthyl splitting esterase during freeze-drying, embedding in poly-ethylene glycol and post fixation. *Acta Histochem.*, **6,** 165–73.

FREISINGER, F. S. 1961. An antibody to urease and its possible relation to gastric ulcer. Ph.D. thesis. Univ. of London.

GAD, A., and SYLVÉN, B. 1969. On the nature of the high iron diamine method for sulfomucins. *J. Histochem. Cytochem.*, **17,** 156–60.

GAHAN, P. B. 1963. Histochemical studies on proliferating cells with special reference to lipids and deoxyribonucleic acid. Ph.D. thesis. Univ. of London.

GAHAN, P. B. 1967. Lysosomes. In *Plant Cell Organelles*. Pridham, J. B., Ed. Academic Press, New York.

GAHAN, P. B. 1972. Ed. *Autoradiography for Biologists*. Academic Press, New York.

GAHAN, P. B., and RAJAN, A. K. 1965. Autoradiography of $^{35}SO_4$ in plant tissue. *Exptl Cell Res.*, **38,** 204–7.

GANOTE, C. E., ROSENTHAL, A. S., MOSES, H. L., and TICE, L. W. 1969. Lead and phosphate as sources of artifact in nucleoside phosphate histochemistry. *J. Histochem. Cytochem.*, **17,** 641–50.

GARRETT, J. R. 1966a. The innervation of salivary glands. 1. Cholinesterase-positive nerves in normal glands of the cat. *J. Roy. micr. Soc.*, **85,** 135–48.

GARRETT, J. R. 1966b. The innervation of salivary glands. 3. The effects of certain experimental procedures on cholinesterase positive nerves in glands of the cat. *J. Roy. micr. Soc.*, **86,** 1–14.

GIACOBINI, E. 1969. Value and limitations of quantitative chemical studies in individual cells. *J. Histochem. Cytochem.*, **17,** 139–55.

GIBBS, H. D. 1927. Phenol tests. II. Nitrous acid tests. The Million and similar tests. Spectrophotometric investigations. *J. Biol. Chem.*, **71,** 445–59.

GLENNER, G. G., WEISSBACH, H., and REDFIELD, B. G. 1960. The histochemical demonstration of enzymatic activity by a nonenzymatic redox reaction. Reduction of tetrazolium salts by indolyl-3-acetaldehyde. *J. Histochem. Cytochem.*, **8,** 258–61.

GLICK, D. 1962. *Quantitative Chemical Techniques of Histo- and Cytochemistry*. Vol. I. Wiley, New York.

GLICK, D. 1963. *Quantitative Chemical Techniques of Histo- and Cytochemistry*. Vol. II. Wiley, New York.

REFERENCES

Glock, G. E., and McLean, P. 1953. Further studies on the properties and assay of glucose 6-phosphate dehydrogenase and 6-phosphogluconate dehydrogenase of rat liver. *Biochem. J.*, **55**, 400–8.

Glock, G. E., and McLean, P. 1954. Levels of enzymes of the direct oxidative pathway of carbohydrate metabolism in mammalian tissues and tumours. *Biochem. J.*, **56**, 171–75.

Glock, G. E., and McLean, P. 1958. Pathways of glucose utilization in mammary tissue. *Proc. Roy. Soc., B*, **149**, 354–62.

Godlewski, H. G. 1960. Application of ethylene-diamine-tetracetic acid (EDTA) in the histochemical method for demonstration of phosphorylase and the branching enzyme. *Bull. l'Acad. Pol. Sci.* (*Sér. biol.*), **8**, 441–44.

Godlewski, H. G. 1962. The distribution of the histochemically demonstrable phosphorylases in various cancer tissues. *Acta Un. Int. Contra Canc.*, **8**, 63–65.

Godlewski, H. G. 1963. Histochemical studies of phosphorylases in precancerous lesions of the uterine cervix and mammary gland. *Fol. Histochem. Cytochem.*, **1**, 231–35.

Godlewski, H. G. 1964. Histochemistry of the glycogen synthetase and phosphorylases in normal and pathologic tissues. *Acta Histochem., Suppl.*, **4**, 30–51.

Goldfischer, S. 1965. The cytochemical demonstration of lysosomal aryl sulfatase activity by light and electron microscopy. *J. Histochem. Cytochem.*, **13**, 520–22.

Goldfischer, S., and Essner, E. 1969. Further observations on the peroxidatic activities of microbodies (peroxisomes). *J. Histochem. Cytochem.*, **17**, 681–85.

Gomori, G. 1952. *Microscopic Histochemistry*. University Press, Chicago.

Gomori, G. 1955. Histochemistry of human esterases. *J. Histochem. Cytochem.*, **3**, 479–84.

Gomori, G. 1956. Histochemical methods for protein-bound sulphydryl and disulphide groups. *Quart. J. micr. Sci.*, **97**, 1–9.

Graham, R. C., and Karnovsky, M. J. 1966. The early stages of absorption of injected horseradish peroxidase in the proximal tubules of mouse kidney: ultrastructural cytochemistry by a new technique. *J. Histochem. Cytochem.*, **14**, 291–302.

Green, D. E. 1959. Structure–function interrelationships in mitochondrial electron transport and oxidative phosphorylation. *Disc. Faraday Soc.*, **27**, 206.

Gunlack, B. G., Neal, G. E., and Williams, D. C. 1966. The demonstration of tetrahydrofolate-dehydrogenase activity in tissues. *Biochem. J.*, **101**, 29–30P.

Hale, C. W. 1946. Histochemical demonstration of acid polysaccharides in animal tissues. *Nature*, **157**, 802.

Hamberger, B., and Norberg, K.-A. 1964. Histochemical demonstration of monoamines in fresh-frozen sections. *J. Histochem. Cytochem.*, **12**, 48.

Handbook of Chemistry and Physics. 1930. By Hodgman, C. D., and Lange, N. A. Chemical Rubber Publishing Company, Cleveland.

Hardonk, M. J. 1968. 5′-nucleotidase. I. Distribution of 5′-nucleotidase in tissues of rat and mouse. *Histochemie*, **12**, 1–17.

Hardonk, M. J., and Boer, H. G. A. de. 1968. 5′-Nucleotidase. III. Determinations of 5′-nucleotidase isoenzymes in tissues of rat and mouse. *Histochemie*, **12**, 29–41.

Hardonk, M. J., and Duijn, P. van. 1964. The mechanism of the Schiff reaction as studied with histochemical model systems. *J. Histochem. Cytochem.*, **12**, 748–51.

Hardonk, M. J., and Koudstaal, J. 1968. 5′-Nucleotidase. II. The significance of 5′-nucleotidase in the metabolism of nucleotides studied by histochemical and biochemical methods. *Histochemie*, **12**, 18–28.

Hart, D. McK. 1966. Hydroxysteroid dehydrogenase activity in normal human placenta from six weeks to forty-two weeks gestation. *J. Endocrinol.*, **35**, 255–62.

Haurowitz, F. 1950. *Chemistry and Biology of Proteins*. Academic Press, New York.

Hayashi, M., Nakajima, Y., and Fishman, W. H. 1964. The cytologic demonstration of β-glucuronidase employing naphthol AS-BI glucuronide and hexazonium pararosanilin; a preliminary report. *J. Histochem. Cytochem.*, **12**, 293–97.

Hayhoe, F. G. J. 1960. *Leukaemia; Research and Clinical Practice*. Churchill, London.

REFERENCES

Hayhoe, F. G. J., and Quaglino, D. 1958. Cytochemical demonstration and measurement of leucocyte alkaline phosphatase activity in normal and pathological states by a modified azo-dye coupling technique. *Brit. J. Haematol.*, **4,** 375–89.

Heilbrunn, L. V. 1943. *An Outline of General Physiology.* 2nd edn. Chaps. 33 and 37. Saunders, Philadelphia.

Heppel, L. A., and Hilmoe, R. J. 1951. Purification and properties of 5-nucleotidase. *J. Biol. Chem.*, **188,** 665–78.

Hewitt, L. F. 1950. *Oxidation–reduction Potentials in Bacteriology and Biochemistry.* 6th edn. Livingstone, Edinburgh.

Hewitt, L. F. 1961. Standard oxidation–reduction potentials of inorganic systems and dyes. In *Biochemists' Handbook.* Long, C., Ed. pp. 85–89. Spon, London.

Höfler, K. 1949. Fluorochromierungsstudien an Pflanzenzellen. In *Beiträge zur Fluoreszenzmikroskopie.* Brautigam, F., and Grabner, A., Eds. Fromme, Vienna.

Holt, S. J. 1952. A new principle for the histochemical localization of hydrolytic enzymes. *Nature,* **169,** 271–73.

Holt, S. J. 1958. Indigogenic staining methods for esterases. In *General Cytochemical Methods.* Danielli, J. F., Ed. Academic Press, New York.

Holt, S. J. 1959. Factors governing the validity of staining methods for enzymes, and their bearing upon the Gomori acid phosphatase technique. *Exptl Cell Res., Suppl.* **7,** 1–27.

Holt, S. J. 1963. Discussion in *Ciba Symp. on 'Lysosomes'.* de Reuck, A. V. S., and Cameron, M. P., Eds. pp. 375–78. Churchill, London.

Holt, S. J., Hobbiger, E. E., and Pawan, G. L. S. 1960. Preservation of integrity of rat tissues for cytochemical staining purposes. *J. Biophys. Biochem. Cytol.*, **7,** 383–86.

Holt, S. J., and O'Sullivan, D. G. 1958. Studies in enzyme cytochemistry. 1. Principles of cytochemical staining methods. *Proc. Roy. Soc., B,* **148,** 465–80.

Holt, S. J., and Sadler, P. W. 1958. Studies in enzyme cytochemistry. III. Relationships between solubility, molecular association and structure in indigoid dyes. *Proc. Roy. Soc., B,* **148,** 495–505.

Holt, S. J., and Withers, R. F. J. 1958. Studies in enzyme cytochemistry. V. An appraisal of indigogenic reactions for esterase localization. *Proc. Roy. Soc., B,* **148,** 520–32.

Holter, H., and Li, S.-O. 1950. Phosphamidase activity of some proteolytic enzymes and rennin. *Acta Chem. Scand.*, **4,** 1321–22.

Holter, H., and Li, S.-O. 1951. Determination and properties of phosphamidase. *Compt. rend. trav. lab. Carlsberg* (*Sér. chim.*), **27,** 393–407.

Hopsu-Havu, V. K., Mäkinen, K. K., and Glenner, G. G. 1966. Formation of bradykinin from kallidin-10 by aminopeptidase B. *Nature, Lond.*, **212,** 1271–72.

Horecker, B. L., and Mehler, A. H. 1955. Carbohydrate metabolism. *Ann. Rev. Biochem.*, **24,** 207–74.

Hotchkiss, R. D. 1948. A microchemical reaction resulting in the staining of polysaccharide structures in fixed tissue preparations. *Arch. Biochem.* **16,** 131–41.

Howe, A. 1959. The distribution of arginine in the pituitary gland of the rat with particular reference to its presence in neurosecretory material. *J. Physiol.*, **149,** 519–25.

Husain, O. A. N., and Cameron, C. B. 1966. Automated methods for cancer screening. *Proc. Roy. Soc. Med.*, **59,** 982–86.

Jacobsen, N. O., and Jørgensen, P. L. 1969. A quantitative biochemical and histochemical study of the lead method for localization of adenosine triphosphate-hydrolyzing enzymes. *J. Histochem. Cytochem.* **17,** 443–53.

Jacobson, W., and Webb, M. 1952. The two types of nucleoproteins during mitosis. *Exptl Cell Res.*, **3,** 163–83.

Jacoby, F., and Martin, B. F. 1949. The histochemical test for alkaline phosphatase. *Nature,* **163,** 875–76.

JARDETSKY, C. D., and GLICK, D. 1956. Studies in histochemistry, XXXVIII. Determination of succinic dehydrogenase in microgram amounts of tissue and its distribution in rat adrenal. *J. Biol. Chem.*, **216**, 283–92.

JEANLOZ, R. W. 1960. The nomenclature of mucopolysaccharides. *Arth. and Rheum.*, **3**, 233–37.

JENSEN, W. A. 1955. The histochemical localization of peroxidase in roots and its induction by indolacetic acid. *Plant Physiol.*, **38**, 426–32.

JOHANSEN, D. A. 1940. *Plant Microtechnique*. McGraw-Hill, New York.

JONES, G. R. N. 1963. Succinic dehydrogenase levels in livers of rats during early feeding of 4-dimethylaminoazobenzene; a reinterpretation of the biochemical data. *Brit. J. Canc.*, **17**, 153–61.

JONES, G. R. N. 1964. Quantitative histochemistry; design and use of a simple microcell for standardized incubation on the slide. *Stain Technol.*, **39**, 155–61.

JONES, G. R. N. 1965. Losses of nitrogenous material occurring from frozen sections of rat liver during incubation in an aqueous medium. *Biochem. J.*, **96**, 10P.

JONES, G. R. N., BITENSKY, L., CHAYEN, J., and CUNNINGHAM, G. J. 1961. Tissue dilution artefact; a re-interpretation of variations in levels of succinic dehydrogenase during chemical carcinogenesis. *Nature*, **191**, 1203.

JONES, G. R. N., MAPLE, A. J., AVES, E. K. CHAYEN, J., and CUNNINGHAM, G. J. 1963. Quantitative histochemistry of succinate dehydrogenase in tissue sections. *Nature*, **197**, 568–70.

JUHLIN, L. 1967. Determination of histamine in small biopsies and histological sections. *Acta Physiol. Scand.*, **71**, 30.

KARMARKAR, S., PEARSE, A. G. E., and SELIGMAN, A. M. 1960. Preparation of nitrotetrazolium salts containing benzothiazole. *J. Org. Chem.*, **25**, 575–78.

KASDON, S. C., FISHMAN, W. H., and HOMBURGER, F. 1950. Beta-glucuronidase studies in women. 2. Cancer of the cervix uteri. *J. Am. Med. Ass.*, **144**, 892–96.

KASDON, S. C., HOMBURGER, F., YORSHIS, E., and FISHMAN, W. H. 1953. Beta-glucuronidase studies in women. VI. Premenopausal vaginal fluid values in relation to invasive cervical cancer. *Surg. Gynec. Obstet.*, **97**, 579–83.

KASDON, S. C., MCGOWAN, J., FISHMAN, W. H., and HOMBURGER, F. 1951. Beta-glucuronidase studies in women. III. Trichomonas vaginalis vaginitis and vaginal fluid enzyme activity. *Am. J. Obstet. Gynec.*, **61**, 647–52.

KASTEN, F. H. 1960. The chemistry of Schiff's reagent. *Int. Rev. Cytol.*, **10**, 1–100.

KAWIAAK, J., SAWICKI, W., and MIKS, B. 1964. Histochemical reaction on aryl sulphatase C activity. *Acta Histochem.*, **19**, 184–90.

KING, J. 1965. *Practical Clinical Enzymology*. Van Nostrand, London.

KIRKBY, W. W. 1965a. The histochemistry of cultured tissues and its use in assessing cellular toxic effects. Ph.D. thesis. Univ. of London.

KIRKBY, W. W. 1965b. The use of histochemical examination of maintenance-cultured tissue in assessing potential cytotoxic substances. *Biochem. J.*, **94**, 24–25P.

KLOTZ, I. M. 1950. The nature of some ion–protein complexes. *Cold Spring Harbour Symp. Quant. Biol.*, **14**, 97–112.

KOELLE, G. B., and FRIEDENWALD, J. S. 1949. A histochemical method for localizing cholinesterase activity. *Proc. Soc. Exptl Biol. Med.*, **70**, 617–22.

KOELLE, G. B., and GROMADZKI, C. G. 1966. Comparison of the gold-thiocholine and gold–thiolacetic acid methods for the histochemical localization of acetylcholinesterase and cholinesterase. *J. Histochem. Cytochem.*, **14**, 443–54.

KOENIG, C. S., and VIAL, J. D. C. 1970. A histochemical study of adenosine triphosphatase in the toad (*Bufo spinulosus*) gastric mucosa. *J. Histochem. Cytochem.*, **18**, 340–53.

KORNBERG, A. 1957. Pathways of enzymatic synthesis of nucleotides and polynucleotides. In *The Chemical Basis of Heredity*. McElvoy, W. D., and Glass, B., Eds. Johns Hopkins, Baltimore.

KORNBERG, H. L. 1961. The breakdown and synthesis of fatty acids. In *Biochemists' Handbook*. Long, C., Ed. pp. 558–63. Spon, London.

KREBS, H. A., and KORNBERG, H. L. 1957. *Energy Transformations in Living Matter*. Springer, Berlin.

REFERENCES

KURNICK, N. B. 1949. Methyl green-pyronin. 1. Basis of selective staining of nucleic acids. *J. Gen. Physiol.*, **33**, 243–64.

KURNICK, N. B. 1950. The quantitative estimation of desoxyribonucleic acid based on methyl green staining. *Exptl Cell Res.*, **1**, 151–58.

KURNICK, N. B., and MIRSKY, A. E. 1949. Methyl green-pyronin. II. Stoichiometry of reaction with nucleic acids. *J. Gen. Physiol.*, **33**, 265–74.

LA COUR, L. F., CHAYEN, J., and GAHAN, P. B. 1958. Evidence for lipid material in chromosomes. *Exptl Cell Res.*, **14**, 469–85.

LAKE, B. D. 1965. The histochemical demonstration of fructose-1-phosphate aldolase and fructose-1-6-diphosphate aldolase, and application of the method to a case of fructose intolerance. *J. Roy. micr. Soc.*, **84**, 489–98.

LAWSON, J. G., and WATKINS, D. K. 1965. Vaginal fluid enzymes in relation to cervical cancer. *J. Obstet. Gynaec. Brit. Comm.*, **72**, 1–8.

LEHNINGER, A. L. 1965. *The Mitochondrion: Molecular Basis of Structure and Function.* Benjamin, New York.

LESSLER, M. A. 1953. The nature and specificity of the Feulgen nucleal reaction. *Int. Rev. Cytol.*, **2**, 231–47.

LEVI, H., and ROGERS, A. W. 1963. On the quantitative evaluation of autoradiographs. *Nat. Fys. Medd. Dan. Vid. Selsk.*, **33**, 5–51.

LEVINE, N. D. 1939. The determination of apparent iso electric points of cell structures by staining at controlled reactions. *Stain Technol.*, **15**, 91.

LILLIE, R. D., and MOWRY, R. W. 1949. Histochemical studies on absorption of iron by tissue reactions. *Bull. Int. A M Mus.*, **30**, 91–94.

LIMA-DE-FARIA, A. 1961. Progress in tritium autoradiography. *Progr. Biophys. Biophys. Chem.*, **12**, 281–317.

LOJDA, Z., PLOEG, M. VAN DER, and DUIJN, P. VAN. 1967. Phosphates of the naphthol AS series in the quantitative determination of alkaline and acid phosphatase activities 'in situ' studied by polyacrylamide membrane model systems and by cytophotometry. *Histochemie*, **11**, 13–32.

LOJDA, Z., VEČEREK, B., and PELICHOVÁ, H. 1964. Some remarks concerning the histochemical detection of acid phosphatase by azo-coupling reactions. *Histochemie*, **3**, 428–54.

LOVELOCK, J. E. 1957. The denaturation of lipid–protein complexes as a cause of damage by freezing. *Proc. Roy. Soc. B.*, **147**, 427–33.

LOVERN, J. A. 1957. *The Chemistry of Lipids of Biochemical Significance.* 2nd edn. Methuen, London.

LOWRY, O. H. 1964. Microanalysis for histochemical purposes. *Proc. 2nd Int. Congr. Histochem. Cytochem.* Springer, Berlin.

LYNCH, R., BITENSKY, L., and CHAYEN, J. 1966. On the possibility of super-cooling in tissues. *J. Roy. micr. Soc.*, **85**, 213–22.

MACLAGAN, N. F. 1964. Diseases of the liver and biliary tract. In *Biochemical Disorders in Human Disease.* Thompson, R. H. S., and King, E. J., Eds. pp. 122–25. Churchill, London.

MAHADEVAN, S., and TAPPEL, A. L. 1967. Arylamidases of rat liver and kidney. *J. Biol. Chem.*, **242**, 2369.

MALMGREN, H., and SYLVÉN, B. 1955. On the chemistry of the thiocholine method of Koelle. *J. Histochem. Cytochem.*, **3**, 441–45.

MANCINI, R. E. 1948. Histochemical study of glycogen in tissues. *Anat. Rec.*, **101**, 149–59.

MARCHESI, V. T., and PALADE, G. E. 1967. The localization of Mg–Na–K-activated adenosine triphosphatase on red cell ghost membranes. *J. Cell Biol.*, **35**, 385.

MARKS, N., DATTA, R. K., and LAJTHA, A. 1968. Partial resolution of brain arylamidases and aminopeptidases. *J. Biol. Chem.*, **243**, 2882–89.

MARRACK, J. 1934. Nature of antibodies. *Nature*, **133**, 292–93.

MASSART, L., PEETERS, G., DE LEY, J., VERCAUTEREN, R., and VON HONCKE, A. 1947. The mechanism of the biochemical activity of acridines. *Experientia*, **3**, 288.

REFERENCES

MATERAZZI, G., and FERRETTI, E. 1970. Histochemical and histophysical investigations on the acetylation blockade of carboxylic groups of polysaccharides. *J. Histochem. Cytochem.*, **18,** 504–9.

MATSCHINSKY, F. M., PASSONNEAU, J. V., and LOWRY, O. H. 1968. Quantitative histochemical analysis of glycolytic intermediates and co-factors with an oil well technique. *J. Histochem. Cytochem.*, **16,** 29–39.

MAYERSBACH, H. V. 1967. Principles and limitations of immunohistochemical methods. *J. Roy. micr. Soc.*, **87,** 295–308.

MAZANOWSKA, A. M., NEUBERGER, A., and TAIT, G. H. 1966. Effect of lipids and organic solvents on the enzymic formation of zinc protoporphyrin and haem. *Biochem. J.*, **98,** 117–27.

MCCABE, M. 1964. The latency of leucine amino-peptidase activity. 2*nd Int. Congr. Histochem. Cytochem.* Springer, Berlin.

MCCABE, M., and CHAYEN, J. 1965. The demonstration of latent particulate aminopeptidase activity, *J. Roy. micr. Soc.*, **84,** 361–71.

MCDONALD, J. K., REILLY, T. J., ZEITMAN, B. B. and ELLIS, S. 1968. Dipeptidyl arylamidase II of the pituitary. *J. Biol. Chem.*, **243,** 2028–37.

MCGEE-RUSSELL, S. M. 1955. A new reagent for the histochemical and chemical detection of calcium. *Nature*, **175,** 301–2.

MCGEE-RUSSELL, S. M. 1958. Histochemical methods for calcium. *J. Histochem. Cytochem.*, **6,** 22–42.

MCLEAN, P., and BROWN, J. 1966. Activities of some enzymes concerned with citrate and glucose metabolism in transplanted rat hepatomas. *Biochem. J.*, **98,** 874–82.

MCLEISH, J. 1959. Comparative microphotometric studies of DNA and arginine in plant nuclei. *Chromosoma*, **10,** 686–710.

MCLEISH, J. L., BELL, L. G. E., LA COUR, L. F., and CHAYEN, J. 1957. The quantitative cytochemical estimation of arginine. *Exptl Cell Res.*, **12,** 120–25.

MCLEISH, J., and SHERRATT, H. S. A. 1958. The use of the Sakaguchi reaction for the cytochemical determination of combined arginine. *Exptl Cell Res.*, **14,** 625–28.

MCMANUS, J. F. A. 1946. Histological demonstration of mucin after periodic acid. *Nature*, **158,** 202.

MELLORS, R. C. 1959. Fluorescent-antibody method. In *Analytical Cytology*. Mellors, R. C., Ed. 2nd edn. pp. 1–67. McGraw-Hill, New York.

MEYER, J., and WEINMAN, J. P. 1953. Phosphamidase content of normal and pathologic tissues of the oral cavity. *J. Histochem. Cytochem.*, **1,** 305–14.

MEYER, J., and WEINMAN, J. P. 1955. A modification of Gomori's method for demonstration of phosphamidase in tissue sections. *J. Histochem. Cytochem.*, **3,** 134–40.

MICHAELIS, L. 1947. The nature of the interaction of nucleic acid and nuclei with basic dyestuffs. *Cold Spring Harbour Symp. Quant. Biol.*, **12,** 131–42.

MOLINE, S. W., and GLENNER, G. G. 1964. Ultrarapid tissue freezing in liquid nitrogen. *J. Histochem. Cytochem.*, **12,** 777–83.

MOSES, H. L., and ROSENTHAL, A. S. 1968. Pitfalls in the use of lead ions for the histochemical localization of nucleoside phosphatases. *J. Histochem. Cytochem.*, **16,** 530–39.

MOSES, H. L., ROSENTHAL, A. S., BEAVER, D. L., and SCHUFFMAN, S. S. 1966. Lead ion and phosphatase histochemistry. II. Effect of adenosine triphosphate hydrolysis by lead ion on the histochemical localization of adenosine triphosphatase activity. *J. Histochem. Cytochem.*, **14,** 702–10.

NACHLAS, M. M., CRAWFORD, D. T., and SELIGMAN, A. M. 1957a. The histochemical demonstration of leucine aminopeptidase. *J. Histochem. Cytochem.*, **5,** 264–78.

NACHLAS, M. M., GOLDSTEIN, T. P., and SELIGMAN, A. M. 1962. An evaluation of aminopeptidase specificity with seven chromogenic substrates. *Arch. Biochem. Biophys.*, **97,** 223–31.

NACHLAS, M. M., MARGULIES, S. I., and SELIGMAN, A. M. 1960. Sites of electron transfer to tetrazolium salts in the succinoxidase system. *J. Biol. Chem.*, **235,** 2739–43.

REFERENCES

NACHLAS, M. M., PRINN, W., and SELIGMAN, A. M. 1956. Quantitative estimation of lyo- and desmoenzymes in tissue sections, with and without fixation. *J. Biophys. Biochem. Cytol.*, **2**, 487–502.

NACHLAS, M. M., and SELIGMAN, A. M. 1949. The histochemical demonstration of esterase. *J. Natl Canc. Inst.*, **9**, 415–25.

NACHLAS, M. M., TSOU, K., SOUZA, E. DE, CHENG, C., and SELIGMAN, A. M. 1957b. Cytochemical demonstration of succinic dehydrogenase by the use of a new *p*-nitrophenyl substituted ditetrazole. *J. Histochem. Cytochem.*, **5**, 420–36.

NACHLAS, M. M., WALKER, D. G., and SELIGMAN, A. M. 1958a. A histochemical method for the demonstration of diphosphopyridine nucleotide diaphorase. *J. Biophys. Biochem. Cytol.*, **4**, 29.

NACHLAS, M. M., WALKER, D. G., and SELIGMAN, A. M. 1958b. The histochemical localization of triphosphopyridine nucleotide diaphorase. *J. Biophys. Biochem. Cytol.*, **4**, 467.

NAIDOO, D., and PRATT, O. E. 1954. The activity and localization of adenosine 5′-phosphatase in the thiamine deficient chicken brain. *Biochim. Biophys. Acta*, **15**, 291–92.

NAUMAN, R. V., WEST, P. W., TROU, F., and GEAKE, G. C. 1960. A spectroscopic study of the Schiff reaction as applied to the quantitative determination of sulfur dioxide. *Anal. Chem.*, **32**, 1307–11.

NIEMI, M., and SYLVÉN, B. 1969. The naphthylamidase reaction as a diagnostic tool for the demonstration of cellular injury and autophagy. *Histochemie*, **18**, 40–46.

NILES, N. R., BITENSKY, L., BRAIMBRIDGE, M. V., and CHAYEN, J. 1966. Histochemical changes related to oxidation and phosphorylation in human heart muscle. *J. Roy. micr. Soc.*, **86**, 159–66.

NILES, N. R., BITENSKY, L., CHAYEN, J., CUNNINGHAM, G. J., and BRAIMBRIDGE, M. V. 1964a. The value of histochemistry in the analysis of myocardial dysfunction. *Lancet*, i, 963–65.

NILES, N. R., CHAYEN, J., CUNNINGHAM, G. J., and BITENSKY, L. 1964b. The histochemical demonstration of adenosine triphosphatase activity in myocardium. *J. Histochem. Cytochem.*, **12**, 740–43.

NOVIKOFF, A. B. 1952. Histochemical demonstration of nuclear enzymes. *Exptl Cell Res., Suppl.* **2**, 123–39.

NOVIKOFF, A. B. 1963a. Discussion in *Ciba Symp. on 'Lysosomes'*. de Reuck, A. V. S., and Cameron, M. P., Eds. pp. 375–78. Churchill, London.

NOVIKOFF, A. B. 1963b. Electron transport enzymes: biochemical and tetrazolium staining studies. In *Histochemistry and Cytochemistry*. Wegmann, R., Ed. Pergamon, Oxford.

NOVIKOFF, A. B. 1967. Enzyme localizations with Wachstein–Meisel procedures: real or artifact. *J. Histochem. Cytochem.*, **15**, 353–54.

NOVIKOFF, A. B. 1970a. Enzyme localizations with Gomori type lead phosphate procedures: real or artifact. *J. Histochem. Cytochem.*, **18**, 366–67.

NOVIKOFF, A. B. 1970b. Their phosphatase controversy: love's labours lost. *J. Histochem. Cytochem.*, **18**, 916–17.

NOVIKOFF, A. B., and GOLDFISCHER, S. 1969. Visualization of peroxisomes (microbodies) and mitochondria with diaminobenzidine. *J. Histochem. Cytochem.*, **17**, 675–80.

NOVIKOFF, A. B., HAUSMAN, D. H., and PODBER, E. 1958. The localization of adenosine triphosphatase in liver: *in situ* staining and cell fractionation studies. *J. Histochem. Cytochem.*, **6**, 61.

ODELL, L. D., and BURT, J. C. 1949. Beta-glucuronidase activity in human female genital cancer. *Canc. Res.*, **9**, 362–65.

ODELL, L. D., BURT, J. C., and BETHEA, R. 1949. β-Glucuronidase activity. *Science*, **109**, 564–65.

ORRHENIUS, S., and ERNSTER, L. 1967. The role of cytochrome *P*-450 in the hydroxylation of lipid-soluble compounds in liver microsomes. *Biochem. J.*, **103**, 3–4P.

OSTROWSKI, K., and BARNARD, E. A. 1961. Application of isotopically-labelled specific inhibitors as a method in enzyme cytochemistry. *Exptl Cell Res.*, **25**, 465–68.

REFERENCES

Ostrowski, K., Barnard, E. A., Stocka, Z., and Darzynkiewicz, Z. 1963. Autoradiographic methods in enzyme cytochemistry. 1. Localization of acetylcholinesterase activity using a ^{3}H-labeled irreversible inhibitor. *Exptl Cell Res.*, **31,** 89–99.

Ostrowski, K., Komender, J., Koščianek, H., and Kwarecki, K. 1962a. Quantitative investigation of the P and N loss in the rat liver when using various media in the 'freeze-substitution' technique. *Experientia,* **18,** 142–46.

Ostrowski, K., Komender, J., Koščianek, H., and Kwarecki, K. 1962b. Quantitative studies on the influence of the temperature applied in freeze-substitution on P, N, and dry mass losses in fixed tissue. *Experientia,* **18,** 227–30.

Ostrowski, K., Komender, J., Koščianek, H., and Kwarecki, K. 1962c. Elution of some substances from the tissues fixed by the 'freeze-substitution' method. *Acta Biochim. Pol.*, **9,** 125–30.

Padykula, H. A., and Herman, E. 1955a. Factors affecting the activity of adenosinetriphosphatase and other phosphatases as measured by histochemical techniques. *J. Histochem. Cytochem.*, **3,** 161–69.

Padykula, H. A., and Herman, E. 1955b. The specificity of the histochemical method for adenosinetriphosphatase. *J. Histochem. Cytochem.*, **3,** 170–83.

Patterson, E. K., Hsiao, S.-H., znd Keppel, A. 1963. Studies on dipeptidases and aminopeptidases. I. Distinction between leucine aminopeptidase and enzymes that hydrolyze L-leucyl-β-naphthylamide. *J. Biol. Chem.*, **238,** 3611.

Patterson, E. K., Hsiao, S.-H., Keppel, A., and Sorof, S. 1965. Studies on dipeptidases and aminopeptidases. II. Zonal electrophoretic separation of rat liver peptidases. *J. Biol. Chem.*, **240,** 710.

Pearse, A. G. E. 1960. *Histochemistry: Theoretical and Applied.* 2nd edn. Churchill, London.

Pearse, A. G. E. 1964. The future of academic and applied histochemistry. *Proc. 2nd Int. Congr. Histochem. Cytochem.* Springer, Berlin.

Pelc, S. R. 1958. Autoradiography as a cytochemical method with special reference to C^{14} and S^{35}. In *General Cytochemical Methods,* Danielli, J. F., Ed. Academic Press, New York.

Plagemann, P. G. W., Gregory, K. F., and Wróblewski, F. 1960a. The electrophoretically distinct forms of mammalian lactic dehydrogenase. I. Distribution of lactic dehydrogenases in rabbit and human tissue. *J. Biol. Chem.*, **235,** 2282–87.

Plagemann, P. G. W., Gregory, K. F., and Wróblewski, F. 1960b. The electrophoretically distinct forms of mammalian lactic dehydrogenase. II. Properties and interrelationships of rabbit and human lactic dehydrogenase isozymes. *J. Biol. Chem.*, **235,** 2288–93.

Ploeg, M. van der, and Duijn, P. van. 1964a. The influence of peroxidases on the DOPA-system. *J. Roy. micr. Soc.*, **83,** 405–14.

Ploeg, M. van der, and Duijn, P. van. 1964b. 5,6-Dihydroxy indole as a substrate in a histochemical peroxidase reaction. *J. Roy. micr. Soc.*, **83,** 415–23.

Pollister, A. W., and Ornstein, L. 1959. The photometric chemical analysis of cells. In *Analytical Cytology,* Mellors, R. C., Ed. 2nd edn. McGraw-Hill, London.

Popják, G. 1961. Biosynthesis of sterols. In *Biochemists' Handbook.* Long, C., Ed. pp. 566–81. Spon, London.

Potter, V. R., Price, J. M., Miller, E. C., and Miller, J. A. 1950. Studies on the intracellular composition of livers from rats fed various aminoazo dyes. III. Effects on succinoxidase and oxaloacetic acid oxidase. *Cancer Res.*, **10,** 28–35.

Pratt, O. E. 1954. Some factors affecting rat brain phosphatase activity in fresh tissue suspensions and in histochemical methods. *Biochim. Biophys. Acta,* **14,** 380–89.

Price, G. R., and Schwartz, S. 1956. Fluorescence microscopy. In *Physical Techniques in Biological Research.* Oster, G., and Pollister, A. W. Eds. Vol. III. Academic Press, New York.

Puchter, H., Meloan, S. N., and Terry, M. S. 1969. On the history and mechanism of alizarin and alizarin red S stains for calcium. *J. Histochem. Cytochem.*, **17,** 110–24.

Richards, B. M. 1960. Redistribution of nuclear proteins during mitosis. In *The Cell Nucleus.* Mitchell, J. S., Ed. p. 138. Butterworth, London.

REFERENCES

RIS, H., and MIRSKY, A. E. 1949. The state of the chromosomes in the interphase nucleus. *J. Gen. Physiol.*, **32,** 489–502.

ROBERTS, D. R. 1966. A rapid staining method giving sharp nuclear definition in frozen sections. *J. med. Lab. Tech.*, **23,** 119–20.

ROBISON, G. A., SCHMIDT, M. J., and SUTHERLAND, E. W. 1970. On the development and properties of the brain adenyl cyclase system. In *Role of Cyclic AMP in Cell Function*, Greengard, P, and Costa, E. Eds. *Advs. in Biochem. Psychopharmacology*, **3,** 11–30. Academic Press, New York.

ROGERS, A. W. 1967. *Techniques of Autoradiography*, Elsevier, Amsterdam.

ROODYN, D. B. 1965. The classification and partial tabulation of enzyme studies on subcellular fractions isolated by differential centrifuging. *Int. Rev. Cytol.*, **18,** 99–190.

ROSENBLATT, D. H., NACHLAS, M. M., and SELIGMAN, A. M. 1958. Synthesis of *m*-methoxy naphthylamines as precursors for chromogenic substrates. *J. Am. Chem. Soc.*, **80,** 2463–65.

ROSENTHAL, A. S., MOSES, H. L., and GANOTE, C. E. 1970. Interpretation of phosphatase cytochemical data. *J. Histochem. Cytochem.*, **18,** 915.

ROSENTHAL, A. S., MOSES, H. L., GANOTE, C. E., and TICE, L. 1969a. The participation of nucleotide in the formation of phosphatase reaction product: a chemical and electron microscope autoradiographic study. *J. Histochem. Cytochem.*, **17,** 839–47.

ROSENTHAL, A. S., MOSES, H. L., TICE, L., and GANOTE, C. E. 1969b. Lead ion and phosphatase histochemistry. III. The effect of lead and adenosine triphosphate concentration on the incorporation of phosphate into fixed tissue. *J. Histochem. Cytochem.*, **17,** 608–12.

RUTENBERG, A. M., COHEN, R. B., and SELIGMAN, A. M. 1952. Histochemical demonstration of arylsulfatase. *Science*, **116,** 539–43.

RUTENBERG, A. M., GOLDBARG, J. A., RUTENBERG, S. H., and LANG, R. I. 1960. The histochemical demonstration of α-d-glucosidase in mammalian tissues. *J. Histochem. Cytochem.*, **8,** 268–72.

RUTENBERG, A. M., RUTENBERG, S. H., MONIS, B., TEAGUE, R., and SELIGMAN, A. M. 1958. Histochemical demonstration of β-d-galactosidase in the rat. *J. Histochem. Cytochem.*, **6,** 122–29.

RUTENBERG, A. M., and SELIGMAN, A. M. 1956. Protein binding of azo dyes by tissue homogenates. *J. Histochem. Cytochem.*, **4,** 17–22.

SAKAGUCHI, S. 1925. Über eine neue Farbreaktion von Protein und Arginin. *J. Biochem.* (*Tokyo*), **5,** 25.

SANDRITTER, W., and KLEINHANS, D. 1964. Über das Trockengewicht, den DNS- und Histonproteingehalt von menschlichen Tumoren. *Zeit. für Krebsforsch.*, **66,** 333–48.

SANDRITTER, W., and FISCHER, R. 1962. Der DNS-Gehalt des normalen Plattenepithels, des Carcinoma in situ und des Invasiven Carcinoms der Portio. *Proc. 1st Int. Congr. Exfol. Cytol.*, Lippincott, Philadelphia, pp. 189–95.

SANDRITTER, W., and KRYGIER, A. 1959. Cytophotometrische Bestimmungen von proteingebundenen Thiolen in der Mitose und Interphase von HeLa-Zellen. *Zeit. für Krebsforsch.*, **62,** 596–610.

SANGER, F. 1945. The free amino groups of insulin. *Biochem. J.*, **39,** 507–15.

SAVARD, K. 1961. Steroid interconversions. In *Biochemists' Handbook*. Long, C., Ed. pp. 581–86. Spon, London.

SCHÜMMELFEDER, N. 1958. Histochemical significance of the polychromatic fluorescence induced in tissues stained with acridine orange. *J. Histochem. Cytochem.*, **6,** 392–93.

SCOTT, J. E. 1970. Histochemistry of Alcian blue. I. Metachromasia of Alcian blue, Astrablau and other cationic phthalocyanin dyes. *Histochemie*, **21,** 277–85.

SCOTT, J. E., and DORLING, J. 1965. Differential staining of acid glycosaminoglycans (mucopolysaccharides) by Alcian blue in salt solutions. *Histochemie*, **5,** 221–33.

SCOTT, J. E., and MOWRY, R. W. 1970. Alcian blue—a consumer's guide. *J. Histochem. Cytochem.*, **18,** 842.

SELIGMAN, A. M. 1963. Developmental histochemistry: the introduction of new methods for light and electron microscopy by the design and preparation of appropriate substrates and reagents. In *Histochemistry and Cytochemistry*. Wegmann, R., Ed. Pergamon, Oxford.

SELIGMAN, A. M., CHAUNCEY, H. H., and NACHLAS, M. M. 1951. Effect of formalin fixation on the activity of five enzymes of rat liver. *Stain Technol.*, **26,** 19–23.

SELIGMAN, A. M., KARNOVSKY, M. J., WASSERKRUG, H. L., and HANKER, J. S. 1968. Nondroplet ultrastructural demonstration of cytochrome oxidase activity with a polymerizing osmiophilic reagent, diaminobenzidine (DAB). *J. Cell Biol.*, **38,** 1–14.

SELIGMAN, A. M., SEITO, T., and PLAPINGER, R. E. 1970a. Some cytochemical correlations between oxidase activity (cytochrome and peroxidase) and chemical structure of bis (phenylenediamines). *Histochemie*, **22,** 85–99.

SELIGMAN, M., UENO, H., HANKER, J., KRAMER, S., WASSERKRUG, H., and SELIGMAN, A. 1966. Cytochemical localization of pancreatic lipase with light and electron microscopy. *Exp. Mol. Path.*, Suppl. **3,** 21–35.

SELIGMAN, A. M., WASSERKRUG, H. L., and PLAPINGER, R. E. 1970b. Comparison of the ultrastructural demonstration of cytochrome oxidase activity with three bis-(phenylenediamines). *Histochemie*, **23,** 63–70.

SERRA, J. A. 1946. Histochemical tests for proteins and amino acids; the characterization of basic proteins. *Stain Technol.*, **21,** 5–18.

SHELLEY, W. B., ÖHMAN, S., and PARNES, H. M. 1968. Mast cell stain for histamine in freeze-dried embedded tissue. *J. Histochem. Cytochem.*, **16,** 433–39.

SHERLOCK, S. 1958. *Diseases of the Liver and Biliary System.* 2nd edn. Univ. Press, Oxford.

SHNITKA, T. K., and SELIGMAN, A. M. 1960. Evidence for the role of esteratic inhibition in producing certain beautiful localization artefacts. *J. Histochem. Cytochem.*, **8,** 344.

SHNITKA, T. K., and SELIGMAN, A. M. 1961. Role of esteratic inhibition on localization of esterase and the simultaneous cytochemical demonstration of inhibitor sensitive and resistant enzyme species. *J. Histochem. Cytochem.*, **9,** 504–27.

SHNITKA, T. K., and SELIGMAN, A. M. 1971. Ultrastructural localization of enzymes. *Ann. Rev. Biochem.*, **40,** 375–96.

SIEKEVITZ, P. 1962. The relationship of cell structure to metabolic activity. In *The Molecular Control of Cellular Activity.* Allen, J. M. Ed. McGraw-Hill, New York.

SILCOX, A. A., POULTER, L. W., BITENSKY, L., and CHAYEN, J. 1965. An examination of some factors affecting histological preservation in frozen sections of unfixed tissue. *J. Roy. micr. Soc.*, **84,** 559–64.

SIMPSON, W. L. 1941. An experimental analysis of the Altmann technic of freezing-drying. *Anat. Rec.*, **80,** 173–89.

SINGER, T. P., and KEARNEY, E. B. 1954. Solubilization, assay and purification of succinic dehydrogenase. *Biochim. Biophys. Acta*, **15,** 151–53.

SKOU, J. C. 1965. Enzymatic basis for active transport of Na^+ and K^+ across cell membranes. *Physiol. Rev.*, **45,** 596.

SLATER, E. C. 1961. Succinic dehydrogenase. In *Biochemists' Handbook*. Long, C., Ed. pp. 369–71. Spoon, London.

SLATER, T. F. 1966. Necrogenic action of carbon tetrachloride in the rat: a speculative mechanism based on activation. *Nature*, **209,** 36–40.

SMITH, E. L. 1951. Proteolytic enzymes. In *The Enzymes*. Sumner, J. B., and Myrbäck, K., Eds. Vol. I. Academic Press, New York.

SMITH, E. L. 1960. Peptide bond cleavage. In *The Enzymes*. Boyer, P. D., Lardy, H., and Myrbäck, K., Eds. Vol. 4. Academic Press, New York.

SMITH, E. L., and SPACKMAN, D. H. 1955. Leucine aminopeptidase. V. Activation, specificity and mechanism of action. *J. Biol. Chem.*, **212,** 271–99.

SNELLMAN, O. 1969. Cathepsin B, the lysosomal thiol proteinase of calf liver. *Biochem. J.*, **114,** 673–78.

REFERENCES

SOVARI, T. E., and STOWARD, P. J. 1970a. Some investigations of the mechanism of the so-called 'methylation' reactions used in mucosubstance histochemistry. I. 'Methylation' with methyl iodide, diazomethane and various organic solvents containing either hydrogen chloride or thionyl chloride. *Histochemie*, **24,** 106–13.

SOVARI, T. E., and STOWARD, P. J. 1970b. Some investigations of the mechanism of the so-called 'methylation' reactions used in mucosubstance histochemistry. II. Histochemical differentiation of lactone and ester groups in 'methylated' mucosaccharides. *Histochemie*, **24,** 114–19.

SPICER, S. S. 1965. Diamine methods for differentiating mucosubstances histochemically. *J. Histochem. Cytochem.*, **13,** 211–34.

STAFFORD, R. O., and ATKINSON, W. B. 1948. Effect of acetone and alcohol fixation and paraffin embedding on activity of acid and alkaline phosphatases in rat tissues. *Science*, **107,** 279–80.

STERNBERG, W. H., FARBER, E., and DUNLAP, C. E. 1956. Histochemical localization of specific oxidative enzymes. II. Localization of diphosphopyridine nucleotide diaphorase and triphosphopyridine nucleotide diaphorase and the succindehydrogenase system in the kidney. *J. Histochem. Cytochem.*, **4,** 266–83.

STOWARD, P. J. 1966. Some comments on the mechanism of the Schiff reaction. *J. Histochem. Cytochem.*, **14,** 681–82.

STOWARD, P. J. 1967a. *J. Roy. micr. Soc.*, **87,** 77–104.

STOWARD, P. J. 1967b, c and d. Studies in fluorescence histochemistry. Parts I, II and III. *J. Roy. micr. Soc.*, **87,** 215, 237, 247.

STRAUS, W. 1959. Rapid cytochemical identification of phagosomes in various tissues of the rat and their differentiation from mitochondria by the peroxidase method. *J. Biophys. Biochem. Cytol.*, **5,** 193–204.

STRAUS, W. 1962. Cytochemical investigation of phagosomes and related structures in cryostat sections of the kidney and liver of rats after intravenous administration of horseradish peroxidase. *Exptl Cell Res.*, **27,** 80–94.

STRUGGER, S. 1938. Die Vitalfarbung des Protoplasmas mit Rhodamin B und 6G. *Protoplasma*, **30,** 85–100.

STUART, J., and SIMPSON, J. S. 1970. Dehydrogenase enzyme cytochemistry of unfixed leucocytes. *J. Clin. Path.*, **23,** 517–21.

STUART, J., SIMPSON, J. S., and MANN, J. R. 1970. Intracellular hydrogen transport systems in acute leukaemia. *Brit. J. Haematol.*, **19,** 739–48.

SWIFT, H. H. 1953. Quantitative aspects of nuclear nucleoproteins. *Int. Rev. Cytol.*, **2,** 1–76.

SWIFT, H. H., and RASCH, E. 1956. Microphotometry with visible light. In *Physical Techniques in Biological Research.* Oster, G., and Pollister, A. W., Eds. Vol. III. Academic Press, New York.

SYLVÉN, B., 1954. Metachromatic dye-substrate interactions. *Quart. J. micr. Sci.*, **95,** 327–58.

SYLVÉN, B. 1968. Studies on the histochemical 'leucine aminopeptidase' reaction. VI. The selective demonstration of cathepsin B activity by means of the naphthylamide reaction. *Histochemie*, **15,** 150–59.

SYLVÉN, B. 1970. Studies on the histochemical 'leucine aminopeptidase' reaction. VII. Control experiments as to the validity and nature of this reaction pattern. *Histochemie*, **24,** 65–84.

SYLVÉN, B., and BOIS, I. 1963. Studies on the histochemical 'leucine aminopeptidase' reaction. II. Chemical and histochemical comparison of the enzymatic and environmental factors involved. *Histochemie*, **3,** 341–53.

SYLVÉN, B., and BOIS-SVENSSON, I. 1964. Studies on the histochemical 'leucine aminopeptidase' reaction. IV. Chemical and histochemical characterization of the intracellular and stromal LNA reactions in solid tumour transplants. *Histochemie*, **4,** 135–49.

SYLVÉN, B., and LIPPI, U. 1965. The suggested lysosomal localization of aminoacylnaphthylamide splitting enzymes. *Exp Cell Res.*, **40,** 145–47.

REFERENCES

SYLVÉN, B., and MALMGREN, H. 1957. The histological distribution of proteinase and peptidase activity in solid mouse tumour transplants. A histochemical study on the enzymic characteristics of different cell types. *Acta radiol., Suppl.* **154,** 1–124.

SYLVÉN, B., and SNELLMAN, O. 1962. A new metal-dependent enzyme present in cathepsin C preparations. *Biochim. Biophys. Acta,* **65,** 350–53.

SYMPOSIUM OF THE INSTITUTE OF BIOLOGY. 1952. Freezing and Drying.

TAKEUCHI, T., and KURIAKI, H. 1953. Histochemical detection of phosphorylase in animal tissues. *J. Histochem. Cytochem.,* **3,** 153–60.

TAPPEL, A. L. 1969. Lysosomal enzymes and other components. In *Lysosomes in Biology and Pathology.* Dingle, J. T., and Fell, H. B., Eds. Vol. II, 207–44. North Holland, Amsterdam.

TAPPEL, A. L., SAWANT, P. L., and SHIBKO, S. 1963. Lysosomes: distribution in animals, hydrolytic capacity and other properties. In *Ciba Symp. on 'Lysosomes'.* de Reuck, A. V. S., and Cameron, M. P., Eds. Churchill, London.

TATA, J. R. 1964. *Biochem. J.,* **90,** 284. Quoted by Roodyn (1965).

TAYLOR, J. H. 1956. Autoradiography at the cellular level. In *Physical Techniques in Biological Research.* Oster, G., and Pollister, A. W., Eds. Vol. III. Academic Press, New York.

TICE, L. W. 1969. Lead–adenosine triphosphate complexes in adenosine triphosphatase histochemistry. *J. Histochem. Cytochem.,* **17,** 85–94.

TORMEY, J. McD. 1966. Significance of the histochemical demonstration of ATPase in epithelia noted for ion transport. *Nature, Lond.,* **210,** 820–22.

UNDERHAY, E., HOLT, S. J., BEAUFAY, H., and DUVE, C. DE. 1956. Intracellular localization of esterase in rat liver. *J. Biophys. Biochem. Cytol.,* **2,** 635–37.

VERCAUTEREN, R. 1950. The structure of desoxyribose nucleic acid in relation to the cytochemical significance of the methyl green-pyronin staining. *Enzymologia,* **14,** 134–40.

VILLEE, C. A. 1962. The role of steroid hormones in the control of metabolic activity. In *The Molecular Control of Cellular Activity.* Allen, J. M., Ed. McGraw-Hill, New York.

VOLK, B. W., and POPPER, H. 1964. Microscopic demonstration of fat in urine and stool by means of fluorescence microscopy. *Am. J. Clin. Path.,* **14,** 234–38.

WACHSTEIN, M. 1946. Alkaline phosphatase activity in normal and abnormal human blood and bone marrow cells. *J. Lab. Clin. Med.,* **31,** 1–17.

WACHSTEIN, M. 1955. Histochemistry of leukocytes. *Ann. N.Y. Acad. Sci.,* **59,** 1052–65.

WACHSTEIN, M., and MEISEL, E. 1952. Histochemical demonstration of 5-nucleotidase activity in cell nuclei. *Science,* **115,** 652–53.

WACHSTEIN, M., and MEISEL, E. 1956. Histochemistry of substrate specific phosphatases at a physiological pH. *J. Histochem. Cytochem.,* **4,** 424–25.

WACHSTEIN, M., and MEISEL, E. 1957. Histochemistry of hepatic phosphatases at a physiologic pH. *Am. J. Clin. Path.,* **27,** 13–23.

WACHSTEIN, M., and MEISEL, E. 1964. Demonstration of peroxidase activity in tissue sections. *J. Histochem. Cytochem.,* **12,** 538–44.

WACHSTEIN, M., MEISEL, E., and NIEDZWIEDZ, A. 1960. Histochemical demonstration of mitochondrial adenosine triphosphatase with the lead–adenosine triphosphate technique. *J. Histochem. Cytochem.,* **8,** 387–88.

WEST, E. S., and TODD, W. R. 1961. *Textbook of Biochemistry.* Macmillan, New York.

WHITE, R. G. 1960. Fluorescent antibody techniques. In *Tools of Biological Research.* Atkins, H. J. B., Ed. 2nd series. Blackwell, Oxford.

WIED, G. L., and BAHR, G. F. 1970. Eds. *Introduction to Quantitative Cytochemistry II.* Academic Press, New York.

WIGGLESWORTH, V. B. 1952. The role of iron in histological staining. *Quart. J. micr. Sci.,* **93,** 105–18.

WILLIGHAGEN, R. G. J., HEUL, R. O. VAN DER, and RIJSSEL, TH. G. VAN. 1963. Enzyme histochemistry of human lung tumours. *J. Path. Bact.,* **85,** 279–80.

WILLIGHAGEN, R. G. J., and PLANTEIJDT, H. T. 1959. Aminopeptidase activity in cancer cells. *Nature,* **183,** 4653–54.

REFERENCES

Wilson, I. B., Hatch, M. A., and Ginsburg, S., 1960. Carbamylation of acetylcholinesterase. *J. Biol. Chem.*, **235**, 2312–15.

Wolman, M. 1955. Problems of fixation in cytology, histology and histochemistry. *Int. Rev. Cytol.*, **4**, 79–102.

Woohsmann, H., and Hartrodt, W. 1964. Der Nachweis einer phosphatempfindlichen Sulfatase mit Naphthol-AS Sulfaten. *Histochemie*, **4**, 336–44.

APPENDIX 1: EFFECT OF FIXATION ON ENZYMES

Type	Time	Temp.	Other treatment	Final activity (% of unfixed)	Reference
		ALKALINE PHOSPHASE			
FORMALIN					
Buffered	15 min	4°C	—	76	Nachlas *et al.*, 1956
Buffered	30 min	4°C	—	71	Nachlas *et al.*, 1956
Buffered	60 min	4°C	—	80	Nachlas *et al.*, 1956
Buffered	2 hr	4°C	—	53	Nachlas *et al.*, 1956
Not specified	2–4 hr	4°C	—	75	Pearse, 1960
10% pH7	2 hr	4°C	—	73	Seligman *et al.*, 1951
10% pH7	24 hr	4°C	—	26	Seligman *et al.*, 1951
Buffered	30 min	25°C	—	52	Nachlas *et al.*, 1956
10% pH7	2 hr	25°C	—	35	Seligman *et al.*, 1951
Buffered	30 min	37°C	—	14	Nachlas *et al.*, 1956
10% pH7	2 hr	37°C	—	13	Seligman *et al.*, 1951
FORMOL–SALINE					
10%	2 hr	—	—	9–23	Emmel, 1946
10%	48 hr	—	—	0–24	Emmel, 1946
ACETONE					
Absolute	15 min	4°C	—	85	Nachlas *et al.*, 1956
Absolute	30 min	4°C	—	95	Nachlas *et al.*, 1956
Absolute	60 min	4°C	—	107	Nachlas *et al.*, 1956
Absolute	2 hr	4°C	—	101	Nachlas *et al.*, 1956
Absolute	24 hr	4°C	—	70	Pearse, 1960
Absolute	28 hr	Cold	—	71	Stafford and Atkinson, 1948
Absolute	50 hr	Cold	—	65	Stafford and Atkinson, 1948
Absolute	72 hr	Cold	—	64	Stafford and Atkinson, 1948
Absolute	30 min	25°C	—	80	Nachlas *et al.*, 1956
Absolute	30 min	37°C	—	79	Nachlas *et al.*, 1956
80%	24 hr	Cold	—	75	Doyle, 1948
Absolute	28–72 hr	Cold	Paraffin 2 hr, 56°C	29	Stafford and Atkinson, 1948
Absolute or 80%	24 hr	—	Paraffin 1 hr, 60°C	28	Doyle, 1948

Type	Time	Temp.	Other treatment	Final activity (% of unfixed)	Reference
ETHANOL					
Absolute	15 min	4°C	—	53	Nachlas *et al.*, 1956
Absolute	30 min	4°C	—	62	Nachlas *et al.*, 1956
Absolute	60 min	4°C	—	57	Nachlas *et al.*, 1956
Absolute	2 hr	4°C	—	66	Nachlas *et al.*, 1956
Absolute	30 min	25°C	—	38	Nachlas *et al.*, 1956
Absolute	30 min	37°C	—	27	Nachlas *et al.*, 1956
Absolute	24 hr	—	—	75	Doyle, 1948
80%	24 hr	—	—	75	Doyle, 1948
80%	28 hr	Cold	—	74	Stafford and Atkinson, 1948
80%	50 hr	Cold	—	88	Stafford and Atkinson, 1948
80%	72 hr	Cold	—	91	Stafford and Atkinson, 1948
Absolute or 80%	24 hr	—	Paraffin 1 hr, 60°C	40	Doyle, 1948
80%	28–72 hr	Cold	Paraffin 2 hr, 56°C	20	Stafford and Atkinson, 1948
FREEZE-DRYING					
—	—	—	—	85	Doyle, 1948
—	—	—	—	87	Berenbom *et al.*, 1952
—	—	—	Paraffin ½ hr, 60°C	29	Berenbom *et al.*, 1952
—	—	—	Polyethylene glycol; then alcohol fixation	90	Fredricsson, 1958

Type	Time	Temp.	Other treatment	Final activity (% of unfixed)	Reference
			5-NUCLEOTIDASE		
FORMALIN					
Not specified	24 hr	Cold	—	40–60	Barka and Anderson 1963
ACETONE					
Absolute	Not specified	Cold	Paraffin 20–40 min, 56°C	47	Novikoff, 1952

Type	Time	Temp.	Other treatment	Final activity (% of unfixed)	Reference
			ACID PHOSPHATASE		
FORMALIN					
Buffered	15 min	4°C	—	67	Nachlas *et al.*, 1956
Buffered	30 min	4°C	—	58	Nachlas *et al.*, 1956
Buffered	60 min	4°C	—	67	Nachlas *et al.*, 1956
Buffered	2 hr	4°C	—	42	Nachlas *et al.*, 1956
10% pH7	2 hr	4°C	—	79	Seligman *et al.*, 1951
Formol–calcium	24 hr	2°C	—	8	Holt, 1959
10% pH7	24 hr	4°C	—	55	Seligman *et al.*, 1951
10% pH7	48 hr	4°C	—	21	Seligman *et al.*, 1951
Buffered	30 min	25°C	—	47	Nachlas *et al.*, 1956
10% pH7	2 hr	25°C	—	40	Seligman *et al.*, 1951
Buffered	30 min	37°C	—	24	Nachlas *et al.*, 1956
10% pH7	2 hr	37°C	—	27	Seligman *et al.*, 1951
Formol–calcium	24 hr	2°C	Washed 12 hr in running water	38–41	Holt, 1959
Formol–calcium	24 hr	2°C	Gum-sucrose 24 hr, 2°C	40–43	Holt, 1959
Formol–calcium	24 hr	2°C	Gum-sucrose 7 days, 2°C	55–60	Holt, 1959
Formol–calcium	24 hr	2°C	Gum-sucrose, 30 days, 2°C	55–57	Holt, 1959
Formol–calcium	24 hr	2°C	Gum-sucrose, 2 yr, 2°C	42–43	Holt, 1959
Formol–calcium	24 hr	2°C	Alcohol 50%, 70%, 90%, then absolute each 30 min; toluene; paraffin, two changes at 45°C; then dewaxed in benzene	1·7–1·8	Holt, 1959
FORMOL–SALINE					
10%	14 hr	—	—	30	Emmel, 1946
10%	41 hr	—	—	20	Emmel, 1946
ACETONE					
Absolute	15 min	4°C	—	80	Nachlas *et al.*, 1956
Absolute	30 min	4°C	—	77	Nachlas *et al.*, 1956
Absolute	60 min	4°C	—	81	Nachlas *et al.*, 1956

Type	Time	Temp.	Other treatment	Final activity (% of unfixed)	Reference
ACETONE—*continued*					
Absolute	2 hr	4°C	—	82	Nachlas *et al.*, 1956
Absolute	7 hr	Cold	—	28	Stafford and Atkinson, 1948
Absolute	24 hr	2°C	—	74–76	Holt, 1959
Absolute	28 hr	Cold	—	20	Stafford and Atkinson, 1948
Absolute	50 hr	Cold	—	22	Stafford and Atkinson, 1948
Absolute	72 hr	Cold	—	40	Stafford and Atkinson, 1948
Absolute	72 hr	Cold	—	25	Doyle, 1948
Absolute	30 min	25°C	—	67	Nachlas *et al.*, 1956
Absolute	30 min	37°C	—	70	Nachlas *et al.*, 1956
Absolute	7–72 hr	Cold	Paraffin 2 hr, 56°C	4	Stafford and Atkinson, 1948
Absolute	24 hr	2°C	Paraffin two changes, 30 min, 45°C, then dewaxed in benzene	20–28	Holt, 1959
ETHANOL					
Absolute	15 min	4°C	—	50	Nachlas *et al.*, 1956
Absolute	30 min	4°C	—	41	Nachlas *et al.*, 1956
Absolute	60 min	4°C	—	50	Nachlas *et al.*, 1956
Absolute	2 hr	4°C	—	38	Nachlas *et al.*, 1956
Absolute or 80%	7 hr	Cold	—	29	Stafford and Atkinson, 1948
Absolute or 80%	24 hr	Cold	—	2	Doyle, 1948
Absolute or 80%	28 hr	Cold	—	18	Stafford and Atkinson, 1948
Absolute or 80%	50 hr	Cold	—	20	Stafford and Atkinson, 1948
Absolute or 80%	72 hr	Cold	—	26	Stafford and Atkinson, 1948
Absolute	30 min	25°C	—	32	Nachlas *et al.*, 1956
Absolute	30 min	37°C	—	22	Nachlas *et al.*, 1956
Absolute or 80%	7–72 hr	Cold	Paraffin 2 hr, 56°C	7	Stafford and Atkinson, 1948
FREEZE-DRYING					
—	—	—	—	85	Doyle, 1948
—	—	—	Paraffin 60°C, ½ hr	13	Berenbom *et al.*, 1952
—	—	—	Paraffin 45°C, 3 min, then dewaxed	98	Holt, 1959

Type	Time	Temp.	Other treatment	Final activity (% of unfixed)	Reference
			β-GLUCURONIDASE		
FORMALIN					
Buffered	15 min	4°C	—	38	Nachlas *et al.*, 1956
Buffered	30 min	4°C	—	23	Nachlas *et al.*, 1956
Buffered	60 min	4°C	—	16	Nachlas *et al.*, 1956
Buffered	2 hr	4°C	—	4	Nachlas *et al.*, 1956
Buffered	30 min	25°C	—	6	Nachlas *et al.*, 1956
Buffered	30 min	37°C	—	2	Nachlas *et al.*, 1956
ACETONE					
Absolute	15 min	4°C	—	76	Nachlas *et al.*, 1956
Absolute	30 min	4°C	—	73	Nachlas *et al.*, 1956
Absolute	60 min	4°C	—	60	Nachlas *et al.*, 1956
Absolute	2 hr	4°C	—	71	Nachlas *et al.*, 1956
Absolute	30 min	25°C	—	80	Nachlas *et al.*, 1956
Absolute	30 min	37°C	—	74	Nachlas *et al.*, 1956
Absolute	24 hr	4°C	Paraffin embedding	0	Pearse, 1960
ETHANOL					
Absolute	15 min	4°C	—	33	Nachlas *et al.*, 1956
Absolute	30 min	4°C	—	26	Nachlas *et al.*, 1956
Absolute	60 min	4°C	—	30	Nachlas *et al.*, 1956
Absolute	2 hr	4°C	—	14	Nachlas *et al.*, 1956
Absolute	30 min	25°C	—	9	Nachlas *et al.*, 1956
Absolute	30 min	37°C	—	1	Nachlas *et al.*, 1956

Type	Time	Temp.	Other treatment	Final activity (% of unfixed)	Reference
			ESTERASE		
FORMALIN					
Buffered	15 min	4°C	—	45	Nachlas *et al.*, 1956
Buffered	30 min	4°C	—	54	Nachlas *et al.*, 1956
Buffered	60 min	4°C	—	43	Nachlas *et al.*, 1956
Buffered	2 hr	4°C	—	15	Nachlas *et al.*, 1956
Formol–calcium	24 hr	2°C	—	20–24	Holt *et al.*, 1960
Buffered	30 min	25°C	—	35	Nachlas *et al.*, 1956
Buffered	30 min	37°C	—	22	Nachlas *et al.*, 1956
Formol–calcium	24 hr	2°C	Washed 12 hr in running water	39–42	Holt *et al.*, 1960
Formol–calcium	24 hr	2°C	Gum-sucrose, 24 hr, 2°C	48–52	Holt *et al.*, 1960
Formol–calcium	24 hr	2°C	Gum–sucrose, 7 days, 2°C	55–62	Holt *et al.*, 1960
Formol–calcium	24 hr	2°C	Gum–sucrose, 30 days, 2°C	50–56	Holt *et al.*, 1960
Formol–calcium	24 hr	2°C	Gum–sucrose, 2 yr, 2°C	33–34	Holt *et al.*, 1960
Formol–calcium	24 hr	2°C	Alcohol 50%, 70%, 90% then absolute each 30 min; two changes of paraffin at 45°C, then dewaxed in benzene	4·6–7	Holt *et al.*, 1960
ACETONE					
Absolute	15 min	4°C	—	61	Nachlas *et al.*, 1956
Absolute	30 min	4°C	—	68	Nachlas *et al.*, 1956
Absolute	60 min	4°C	—	62	Nachlas *et al.*, 1956
Absolute	2 hr	4°C	—	59	Nachlas *et al.*, 1956
Absolute	24 hr	2°C	—	62–70	Holt *et al.*, 1960
Absolute	24 hr	2°C	Toluene; two changes of paraffin at 45°C then dewaxed in benzene	26–30	Holt *et al.*, 1960

Type	Time	Temp.	Other treatment	Final activity (% of unfixed)	Reference
ETHANOL					
Absolute	24 hr	Cold	—	0	Nachlas and Seligman, 1949
FREEZE-DRYING					
—	—	—	Paraffin 5 min, 45°C, benzene, four changes, 30 min each at room temp	99	Holt *et al.*, 1960
—	—	—	—	91	Berenbom *et al.*, 1952
—	—	—	Paraffin 60°C ½ hr	10	Berenbom *et al.*, 1952

Type	Time	Temp.	Other treatment	Final activity (% of unfixed)	Reference
			LEUCINE AMINOPEPTIDASE		
FORMALIN					
Buffered	15 min	4°C	—	90	Nachlas *et al.*, 1956
Buffered	30 min	4°C	—	82	Nachlas *et al.*, 1956
Buffered	60 min	4°C	—	85	Nachlas *et al.*, 1956
Buffered	2 hr	4°C	—	54	Nachlas *et al.*, 1956
Buffered	30 min	25°C	—	63	Nachlas *et al.*, 1956
Buffered	30 min	37°C	—	14	Nachlas *et al.*, 1956
ACETONE					
Absolute	15 min	4°C	—	90	Nachlas *et al.*, 1956
Absolute	30 min	4°C	—	76	Nachlas *et al.*, 1956
Absolute	60 min	4°C	—	84	Nachlas *et al.*, 1956
Absolute	2 hr	4°C	—	79	Nachlas *et al.*, 1956
Absolute	30 min	25°C	—	80	Nachlas *et al.*, 1956
Absolute	30 min	37°C	—	77	Nachlas *et al.*, 1956
ETHANOL					
Absolute	15 min	4°C	—	0	Nachlas *et al.*, 1956

Enzyme	Fixation	Final activity	Reference
OTHER ENZYMES			
DPNH–tetrazolium reductase	Cold formol–calcium overnight	5–25	Novikoff, 1963
TPNH–tetrazolium reductase	Cold formol–calcium overnight	4–10	Novikoff, 1963
DPNH–cytochrome *c* reductase	Cold formol–calcium overnight	1–6	Novikoff, 1963
Cytochrome oxidase	Freeze-drying only	56	Berenbom *et al.*, 1952
Cytochrome oxidase	Freeze-drying, then embedded in paraffin ½ hr, 60°C	0	Berenbom *et al.*, 1952
Cytochrome oxidase	Formalin, 2–4 hr 4°C	0	Pearse, 1960
Cytochrome oxidase	Acetone 24 hr, 4°C	traces	Pearse, 1960
Cytochrome oxidase	Cold acetone + embedding	0	Pearse, 1960
Succinic dehydrogenase	Formalin 2–4 hr, 4°C	0	Pearse, 1960
Succinic dehydrogenase	Acetone 24 hr, 4°C	60	Pearse, 1960
Succinic dehydrogenase	Cold acetone + embedding	0	Pearse, 1960
Succinoxidase	Freeze-drying only	37	Berenbom *et al.*, 1952
Succinoxidase	Freeze-drying, then embedded in paraffin ½ hr, 60°C	0	Berenbom *et al.*, 1952

APPENDIX 1

EFFECTS OF GLUTARALDEHYDE FIXATION

(*a*) *Tissue:* Skeletal muscle mince (Anderson, 1967).
Fixative: 4% glutaraldehyde or 4% formaldehyde in 0·135M Sörensen buffer, pH 7·3.
Conditions: 5 g minced muscle in 100 ml test solution for 1 hr at 4°C then washed three times in distilled water.

Enzyme	*Location*	*Final activity (% of unfixed) in:* GLUTARALDEHYDE *Distilled*	*Charcoal purified*	*Stock*	FORMALDEHYDE
Lactate dehydrogenase	Supernatant	10	9	8	20
Lactate dehydrogenase	Sediment	12	5	4	12
α-Hydroxybutyrate dehydrogenase	Supernatant	30	28	20	40
Isocitrate dehydrogenase	Supernatant	18	15	12	20
Aspartate aminotransferase	Supernatant	32	20	12	53
Alanine amino transferase	Supernatant	48	40	28	40
Creatine phosphokinase	Supernatant	82	55	32	90
Creatine phosphokinase	Sediment	12	5	4	12
Cholinesterase	Supernatant	72	50	24	70
Cholinesterase	Sediment	75	45	20	65
Acid phosphatase	Supernatant	40	25	18	42
Acid phosphatase	Sediment	40	35	20	50

(*b*) *Tissue:* Rat liver homogenate in 0·25M sucrose (Fahimi and Drochmans, 1968).
Fixative: 5% glutaraldehyde in water.
Conditions: 4 ml homogenate and 1 ml fixative for 30 min at 4°C.
Enzyme: Acid phosphatase.

Type of glutaraldehyde	*Final activity (% of unfixed)* *Free*	*Total*
Distilled	43	33
Distilled (heated)	44	33
Union carbide	38	28
Fluka	33	24
Fisher (biological grade)	16	13

(*c*) *Tissue:* Rat kidney (Arborgh, Ericsson and Helminen, 1971).
Fixative: 1·5% glutaraldehyde (distilled) in Sörensen phosphate buffer; pH 7·4.
Conditions: On anaesthetized animals, kidney perfused for 30 sec with isotonic saline, and then with fixative for 5 min at 20°C. Then small slices (1mm thick) immersed at 0–4°C washed with 0·1M Tris–maleate buffer, pH 7·4.

Enzyme	*Time of fixation*	*Time of washing*	*Final activity* (% of unfixed)
Acid phosphatase	2 min perfusion	Perfusion rinse	30
Acid phosphatase	5 min perfusion	Perfusion rinse	12
Acid phosphatase	5 min perfusion 1 hr immersion	—	10
Acid phosphatase	5 min perfusion 24 hr immersion	24 hr	10
Acid phosphatase	5 min perfusion 24 hr immersion	7 days	12
Aryl sulphatase	5 min perfusion	Perfusion rinse	30
Aryl sulphatase	5 min perfusion 24 hr immersion	—	30
Aryl sulphatase	5 min perfusion 24 hr immersion	24 hr	53
Aryl sulphatase	5 min perfusion 24 hr immersion	7 days	60

(*d*) *Enzyme:* Adenosine triphosphatase

Type	*Tissue*	*Fixative*	*Time*	*Temp.*	*Final activity* (% of unfixed)	*Reference*
Na^+–K^+ activated	Kidney membrane fractions	Glutaraldehyde	Brief	0°C	0	Moses *et al.*, 1966
Na^+–K^+ activated	Kidney membrane fractions	Formaldehyde	Brief	0°C	0	Moses *et al.*, 1966
Na^+–K^+ activated	Duck salt gland or teleost gill	Buffered glutaraldehyde 0·5%, pH 7·2	10 min	4°C	< 10	Ernst and Philpott, 1970
Na^+–K^+ activated	Duck salt gland or teleost gill	Buffered formaldehyde 2%, pH 7·2	30 min	4°C	100*	Ernst and Philpott, 1970
Na^+–K^+ activated	Duck salt gland or teleost gill	Buffered formaldehyde 2%, pH 7·2	90 min	4°C	65	Ernst and Philpott, 1970
Mg^+ activated	Duck salt gland or teleost gill	Buffered glutaraldehyde 0·5%, pH 7·2	60 min	4°C	15	Ernst and Philpott, 1970
Mg^+ activated	Duck salt gland or teleost gill	Buffered formaldehyde 2%, pH 7·2	60 min	4°C	25–30*	Ernst and Philpott, 1970
Mg^+ activated	Toad gastric mucosa	Buffered glutaraldehyde 6·25%, pH 7·4	2 hour	—	35	Koenig and Vial, 1970

* It is difficult to reconcile these results.

APPENDIX 2

SOURCE LIST OF CHEMICALS USED IN THE HISTOCHEMICAL TECHNIQUES

Chemical	*Specification*	*Source* *U.K.*	*U.S.A.*
Acetic anhydride	Laboratory reagent	B.D.H.	J. T. Baker
Acetonitrile, see Methyl cyanide			
Acetyl thiocholine iodide	—	Koch Lights	Sigma
Acridine orange	Revector†	H. & W.	MCB
Adenosine-5-diphosphate	—	Boehringer	Sigma
Adenosine-5-monophosphate	—	Boehringer	Calbiochem
Adenosine-5-triphosphate	—	Boehringer	Sigma
Adrenaline hydrogen tartrate	—	B.D.H.	Fisher Scientific Co.
Adrenaline (buffered)	—		
Alcian blue	Revector†	H. & W.	Edward Gurr
p-Aminodiphenylamine	Laboratory reagent	B.D.H.	Sigma
2-Amino,2(hydroxy-methyl) propane-1,3-diol	Laboratory reagent	B.D.H.	Sigma
8-Amino-1-naphthyl 3-6-disulphonic acid	Mono sodium salt	H. & W.	Fisher
Aniline	ANALAR	B.D.H.	Sigma
Ascorbic acid	Ba salt	B.D.H.	Sigma
Ascorbic acid	Na salt	Halewood Chemicals	Sigma
Benzoyl chloride	Laboratory reagent	B.D.H.	J. T. Baker
Benzo-α-pyrene	Practical grade	Sigma	Fisher Scientific Co.
5-Bromoindoxyl acetate	Σ grade	Sigma	Sigma
Caffeine		H. & W.	Sigma
Calcium chloride	Dried ($2H_2O$)‡	H. & W.	J. T. Baker
Carmine	Standard stain	B.D.H.	Edward Gurr
Celloidin		H. & W.	Edward Gurr
Charcoal	Activated	H. & W.	Scientific Supplies
p-Chloroanilido-phosphonic acid	Grade 2	Sigma	Sigma
Chrome alum, see Chromic potassium sulphate			
Chromic potassium sulphate		H. & W.	J. T. Baker
Collagen polypeptides	5115, 5163	Sigma	Sigma
Cysteine HCl	L-isomer	H. & W.	Sigma
Cystine	L-isomer	H. & W.	MCB
DePeX		H. & W.	(e.g. Technicon or Permount)
o-Dianisidine-tetrazotized	Practical grade	Sigma	Sigma
Dichloro-1-naphthyl	—	Eastman Kodak	Fisher Scientific Co.

Chemical	Specification	Source U.K.	Source U.S.A.
Dimethyl formamide, see Form-dimethylamide			
Dinitro-fluoro-benzene, see 1-Fluoro 2-4-dinitrobenzene			Sigma
Dinitrophenol	G.P.R.	H. & W.	Sigma
DPN, see NAD			Sigma
DPNH, see $NADH_2$			Sigma
Dry ice	'Cardice'	Distillers Co.	local
E.D.T.A., see Sequestric acid			J. T. Baker
Eosin yellowish	Water soluble	H. & W.	Allied Chem
Euparal	—	H. & W.	A. J. Thomas
Farrants' medium	—	H. & W.	A. J. Thomas
Fast Blue B salt		Sigma	Sigma
Fast Dark Blue R salt		Sigma	Sigma
Fast Green	F.C.F.	B.D.H.	Allied Chem
1-Fluoro-2,4-dinitrobenzene		H. & W.	Sigma
Formalin = 40% formaldehyde (technical)		B.D.H.	
Form-dimethylamide	G.P.R.	B.D.H.	Sigma
Fuchsin (acid)	Microscopical stain	B.D.H.	Allied Chem
Glucose-6-phosphate	Na salt	Boehringer	Calbiochem
L-Glutamic acid	Na salt (puriss.)	Koch Lights	Sigma
Glutathione	Reduced form	Sigma	Sigma
DLαGlycerophosphate	Na salt	Sigma	Sigma
Glycogen	Rabbit liver	B.D.H.	Sigma
Glycyl glycine	—	B.D.H.	Sigma
Guaiacol	—	B.D.H.	Sigma
H-acid, see 8-Amino-1-naphthol-3, 6-disuphonic acid			Fisher
Haemalum (Mayers')	Revector†	H. & W.	MCB
Haematoxylin	Standard stain	B.D.H.	Allied Chem
n-Heptanol	Laboratory reagent	B.D.H.	
Hexane	Free from aromatic hydrocarbons	B.D.H.	MCB
β-Hydroxy-butyric acid	DL isomer. Na salt	Sigma	Sigma
1 Hydroxy-2-naphthoic acid		H. & W.	Sigma
Insulin		Boots	Sigma
L-Leucyl-β-naphthylamide		Sigma	MCB
Mercuric chloride	G.P.R.	H. & W.	
Mercuric sulphate	M.A.R.	B.D.H.	J. T. Baker
Mercury orange		Eastman Kodak	Sigma
Methyl acetonitrile (methyl cyanide)			Fisher Scientific Co.
Methyl cyanide	Laboratory reagent	B.D.H.	
Methyl green	Revector†	H. & W.	Allied Chem
Methyl violet	Technical dye	H. & W.	Allied Chem

Chemical	Specification	Source U.K.	Source U.S.A.
'Milton'	Proprietary brand	Boots	
NAD	Free acid	Boehringer	Sigma
$NADH_2$	Na salt	Boehringer	Sigma
NADP	Na salt	Boehringer	Sigma
$NADPH_2$	Na salt	Boehringer	Calbiochem
α-Naphthol	Grade II	Sigma	Sigma
Naphthol-AS-BI-glucuronide	A. grade	Sigma	Calbiochem
Naphthol AS-BI phosphate		Sigma	Sigma
Naphthol AS-TR phosphate		Sigma	Sigma
α-Naphthyl acetate	Laboratory reagent	H. & W.	Fisher
α-Naphthyl phosphate	Na salt	Sigma	Sigma
Neotetrazolium chloride		U.C.P. Serva§	Sigma
Nitro-blue tetrazolium	Grade III	Sigma	Sigma
Oil red O	Revector†	H. & W.	Allied Chem
Pararosaniline HCl	Technical dye	H. & W.	Sigma
Periodic acid	G.P.R.	H. & W.	G. F. Smith
Phenazine methosulphate		Sigma	Sigma
p-Phenylene diamine, see *p*-Amino-diphenylamine			Sigma
6-Phosphogluconic acid	Tri Na salt	Wessex	Sigma
Polypeptides, see Collagen polypeptides			
Polyvinyl alcohol	MO5/140, BO5/140	B.B. & S.B.	MCB
Pyronine G.	Revector†	H. & W.	Allied Chem
Ribonuclease	From bovine pancreas	Sigma	Sigma
Schiff's reagent	For aldehydes	H. & W.	Allied Chem
Sequestric acid	G.P.R.	H. & W.	J. T. Baker
Sodium β-glycerophosphate		B.D.H.	Sigma
Sodium lactate	Solution	B.D.H.	
Sodium succinate		H. & W.	MCB
Tetra-azotized dianisidine, see Dianisidine-tetrazotized			Sigma
sym-Tetrachloroethane	G.P.R.	H. & W.	J. T. Baker
Toluidine blue	Revector†	H. & W.	Allied Chem
TPN, see NADP			Sigma
TPNH, see $NADPH_2$			Sigma
Tris, see 2-Amino-2-(hydroxymethyl) propane 1,3-diol			Sigma
Tryptamine HCl		B.D.H.	Sigma
Tween 40.60.80	G.P.R.	H. & W.	Sigma

† Revector brand; or highest quality.

‡ Different sources contain different amounts of water. Care must be taken to ensure that, e.g. a 1% solution contains 1% of calcium chloride and not unknown amounts of water.

§ For highly purified neotetrazolium, particularly for quantitative studies.

APPENDIX 2

ADDRESSES OF BRITISH SUPPLIERS

B.B. & S.B.	Agent: The Cytochemical Company, 42 Wychwood Avenue, Edgware, Middlesex HA8 6TH
B.D.H.	British Drug Houses Ltd., Poole, Dorset BH12 ANN
Boehringer	Boehringer Corp. London Ltd., Bilton House, 54/58 Uxbridge Road, Ealing, London W5 2TZ
Boots	Any retail chemist
Calbiochem	Calbiochem Ltd., 10 Wyndham Place, London W1H 1AS
Distillers Co.	Distillers Co. Ltd. ('phone Reigate 47611 to find your local branch)
Eastman Kodak	Kodak Ltd., Kirkby, Liverpool
Gurr, G. T.	George T. Gurr, Lane End Road, High Wycombe, Buckinghamshire
Halewood Chemicals	Halewood Chemicals Ltd., Stanwell Moor, Staines, Middlesex
H. & W.	Hopkin & Williams Ltd., P.O. Box 1, Romford, Essex RM1 1HA
Koch Lights	Koch–Light Laboratories Ltd., Colnbrook, Buckinghamshire
Serva	Agent: Micro-Bio Laboratories Ltd., 46 Pembridge Road, London W11
Sigma	Sigma Chemical Company Ltd., Norbiton Station Yard, Kingston-upon-Thames, Surrey KT2 7BH
Wessex	Wessex Biochemicals Ltd., Castle Road, Bournemouth BH9 1PH

ADDRESSES OF AMERICAN SUPPLIERS

J. T. Baker Chemical Company,
122 East 42nd Street,
New York,
New York

Sigma Chemical Company,
3500 DeKalb Street,
St. Louis,
Missouri 63118

Matheson Colman & Bell,
2909 Highland Avenue,
Norwood,
Ohio 45212

Calbiochem,
Box 54282,
Los Angeles,
California 90054

Fisher Scientific Company/USA,
633 Greenwich Street,
New York
New York 10014

Fisher Scientific Company/Canada,
8505 Devonshire Road,
Montreal,
Quebec

Scientific Supplies,
3745 Bay Shore Blvd.,
Brisbane,
California

Allied Chemical & Dye Corp.,
National Aniline Division
40 Rector Street,
New York,
New York

G. Frederick Smith Chemical Company,
Columbus,
Ohio

Arthur J. Thomas Company,
Philadelphia

CRYOSTAT MICROTOME

Manufacturer and supplier: Bright Instrument Co. Ltd., Clifton Road, St. Peter's Road, Huntingdon, England

APPENDIX 3

BUFFERS

0·1M PHOSPHATE BUFFER, pH 5·6–8·0

100 ml Mixtures of

A. 0·1M Disodium hydrogen orthophosphate, anhydrous. MW = 141·96
B. 0·1M Potassium dihydrogen orthophosphate. MW = 136·09

pH	ml 0·1M Na_2HPO_4	ml 0·1M KH_2PO_4
5·6	5·5	94·5
5·8	8·0	92·0
6·0	12·0	88·0
6·2	18·0	82·0
6·4	28·0	72·0
6·6	38·0	62·0
6·8	50·0	50·0
7·0	62·0	38·0
7·2	72·0	28·0
7·4	82·0	18·0
7·6	88·0	12·0
7·8	92·0	8·0
8·0	94·5	5·5

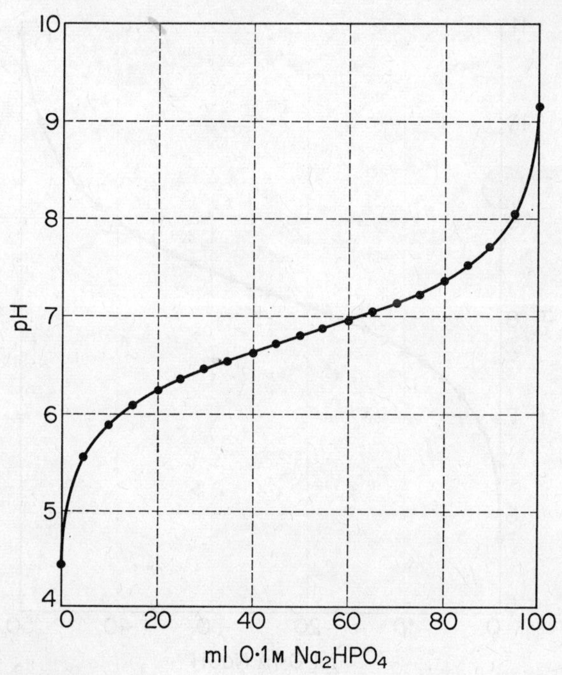

0·05M GLYCYL GLYCINE BUFFER, pH 7·0–9·6

50 ml 0·1M Glycyl glycine (MW = 132·12) + x ml 0·1M sodium hydroxide (MW = 40·00) made up to 100 ml with distilled water.

pH	x ml 0·1M NaOH
7·0	2·5
7·2	4·0
7·4	6·0
7·6	9·0
7·8	12·5
8·0	17·0
8·2	23·0
8·4	28·0
8·6	33·0
8·8	37·5
9·0	40·5
9·2	43·0
9·4	45·0
9·6	46·0

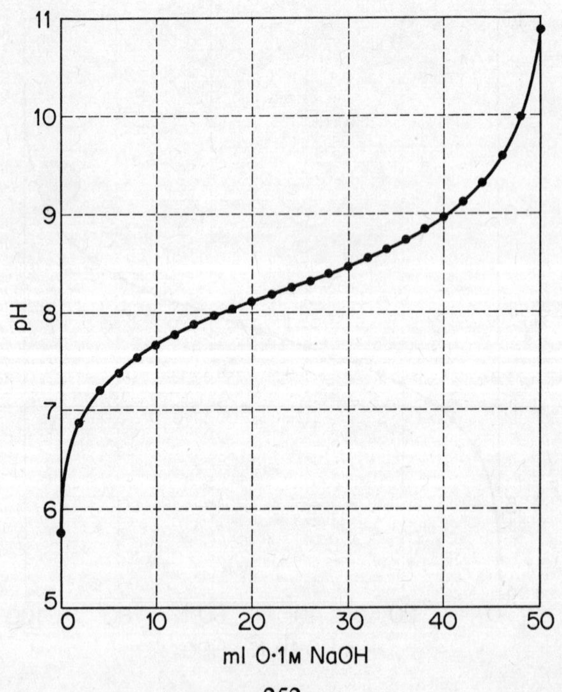

APPENDIX 3

0·05M TRIS BUFFER, pH 7·0–9·2

25 ml 0·2M Tris-(hydroxymethyl)-amino-methane (MW = 121·14) + x ml 0·1 N HCl, made up to 100 ml with distilled water.

pH	x ml 0·1 N HCl
7·0	45·5
7·2	44·0
7·4	42·0
7·6	38·5
7·8	34·5
8·0	29·0
8·2	23·0
8·4	17·5
8·6	13·0
8·8	9·0
9·0	6·5
9·2	4·0

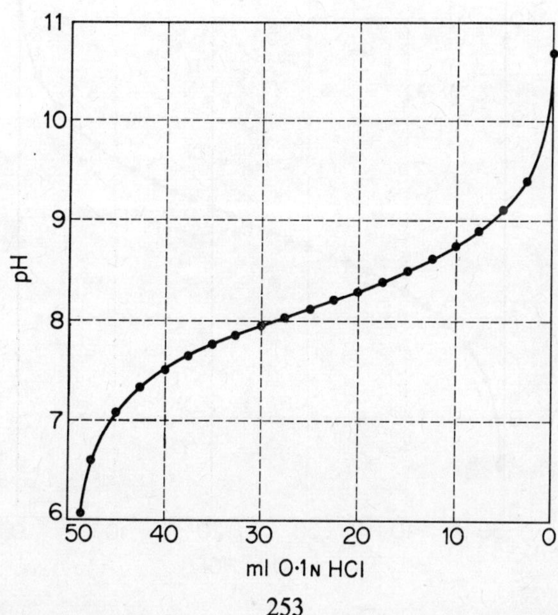

0·05M BARBITONE BUFFER, pH 6·8–9·0

50 ml 0·1M Barbitone-sodium (sodium 5,5-diethylbarbiturate) (MW = 206·18) + x ml 0·1 N HCl, made up to 100 ml with distilled water.

pH	x ml 0·1 N HCl
6·8	45·5
7·0	43·5
7·2	41·0
7·4	37·0
7·6	32·5
7·8	27·5
8·0	22·5
8·2	17·5
8·4	13·0
8·6	9·5
8·8	7·0
9·0	5·0

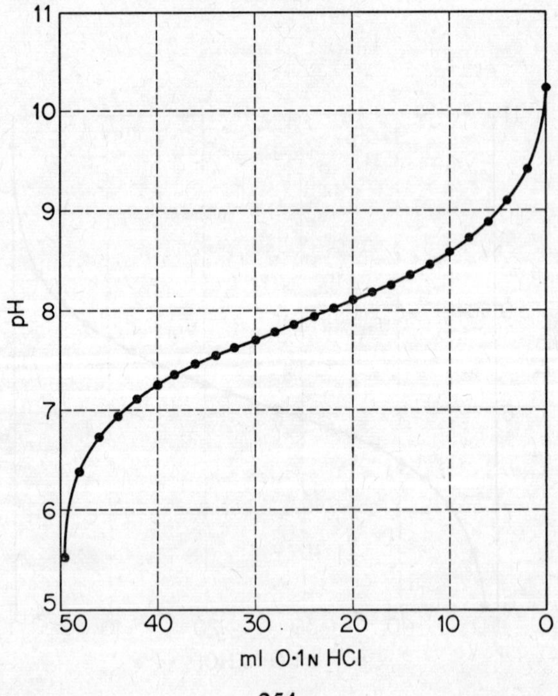

McILVAINE BUFFER, pH 2·2–7·8

100 ml mixtures of 0·2M disodium hydrogen orthophosphate, anhydrous, (MW = 141·96) and 0·1M citric acid (MW = 210·14)

pH	ml 0·1M Citric acid	ml 0·2M Na_2HPO_4
2·2	98·0	2·0
2·4	93·5	6·5
2·6	89·0	11·0
2·8	84·0	16·0
3·0	79·0	21·0
3·2	75·0	25·0
3·4	71·0	29·0
3·6	67·5	32·5
3·8	64·0	36·0
4·0	60·5	39·5
4·2	58·0	42·0
4·4	55·5	44·5
4·6	52·5	47·5
4·8	50·0	50·0
5·0	48·0	52·0
5·2	46·0	54·0
5·4	44·0	56·0
5·6	42·0	58·0
5·8	40·0	60·0
6·0	37·0	63·0
6·2	34·0	66·0
6·4	31·0	69·0
6·6	27·5	72·5
6·8	23·0	77·0
7·0	18·0	82·0
7·2	13·5	86·5
7·4	10·0	90·0
7·6	7·0	93·0
7·8	5·0	95·0

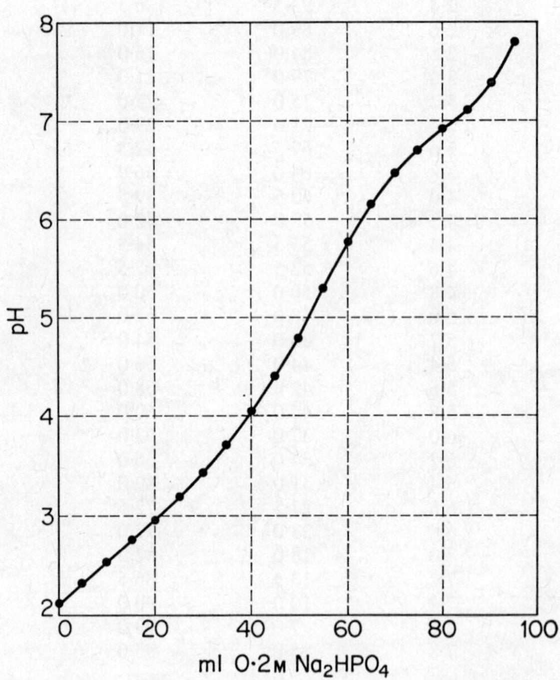

8
7
6
5
4
3
2
pH
0
20
40
60
80
100
ml 0·2M Na2HPO4

0·2M ACETATE BUFFER, pH 3·6–5·6

100 ml mixtures of:

A. 0·2M Sodium acetate, anhydrous MW = 82·03
B. 0·2M Acetic acid MW = 60·00

pH	ml 0·2M Sodium acetate	ml 0·2M Acetic acid
3·6	8·0	92·0
3·8	13·0	87·0
4·0	20·0	80·0
4·2	30·0	70·0
4·4	40·0	60·0
4·6	50·0	50·0
4·8	60·0	40·0
5·0	70·0	30·0
5·2	80·0	20·0
5·4	86·0	14·0
5·6	90·0	10·0

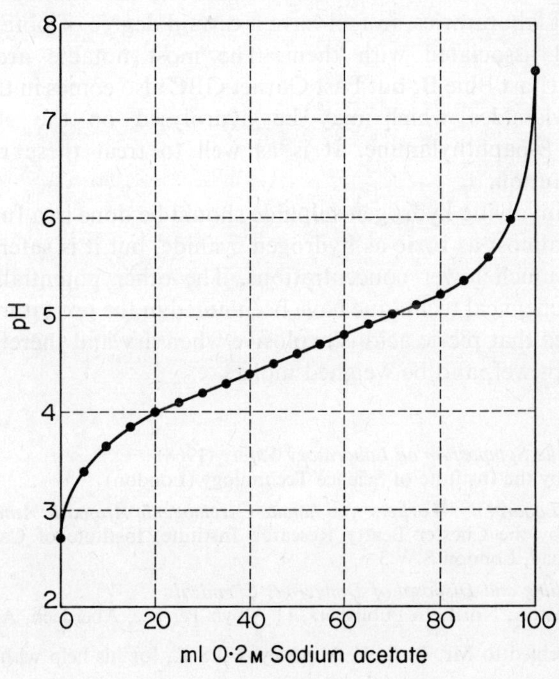

APPENDIX 4

SAFETY*

The histochemical laboratory is no more dangerous than most chemical or biochemical laboratories. As Libman (1968, I.S.T. Symposium) pointed out, 'As in all other types of activity, safety in the handling of chemicals depends on a responsible attitude, knowledge and foresight'. The former attitude is essential in order to maintain a sensible balance between such excessive caution as will banish 'dangerous' chemicals from the laboratory and stultify research, and such foolish disregard of laboratory hazards that is a danger to oneself and to one's colleagues. As regards knowledge and foresight we can only recommend some articles on this subject, and advise on the most obvious dangers. In general this is done in our book in the relevant places, for example when discussing the use of benzpyrene or acetonitrile. And, of course, it should be realized that inhibitors of enzyme reactions in sections are also inhibitors of enzymic biochemistry in the body; cyanide is the obvious example. It is preferable to keep such reagents in a separate compartment in the laboratory, and to have warning labels on the bottles.

It may not be realized that some substances, in common use in some histochemical laboratories, in fact have a certain degree of potential carcinogenic hazard associated with them. The most notable are: benzidine, *o*-dianisidine (Fast Blue B; but Fast Garnet GBC also comes in this category); and naphthylamides which may be hydrolysed on the skin to yield carcinogenic β-naphthylamine. It is as well to treat these reagents with reasonable caution.

Reactions involving hydrogen sulphide should be done in a fume cupboard because it is almost as toxic as hydrogen cyanide, but it is safer because it is nauseous at much lower concentrations. The other potentially dangerous steps in histochemical technique have been noted in the text; it may, however, be emphasized that picric acid is explosive when dry and therefore it should always be kept wet, and be weighed moist.

BIBLIOGRAPHY

Proceedings of the Symposium on Laboratory Safety (1968)
Published by the Institute of Science Technology (London)

Precautions for Laboratory Workers who handle Carcinogenic Aromatic Amines (1966)
published by the Chester Beatty Research Institute, Institute of Cancer Research, Fulham Road, London S.W.3

The Care, Handling and Disposal of Dangerous Chemicals
by P. J. Gaston; Northern Publishers, 11 Albyn Terrace, Aberdeen, AB9 1SE

* We are indebted to Mr. B. C. V. Mitchley, F.I.S.T., for his help with this appendix.

INDEX

(Unless otherwise indicated, the page number of methods is given in **bold** type)

INDEX